Music Production Cultures

Music Production Cultures draws on interviews with international educators, surveys completed by students of music production from around the globe, doctoral research findings and contextualised career experiences from the author as a celebrated music producer to explore how effective learning environments can be created for popular music production in higher education.

Acknowledging the musical, technological and social diversity in global popular music production practice, this book highlights the integral elements that educators and their institutions must consider in order to provide high-quality and relevant education for the students of today and into the future. Offering concepts, approaches and practices to be integrated into diverse music production pedagogical frameworks in higher education, this book considers the pedagogical approaches and goals that bridge music production education to the industry, using examples and insights from international educators throughout as well as lesson plan examples for instructors.

Music Production Cultures develops a foundation of practice to inform teachers designing equitable, diverse and inclusive pedagogies that are dependent on the musical, cultural and social influences of their students. This is an invaluable resource for educators and researchers in the area of audio education looking to develop their pedagogical strategies.

Brendan Anthony has an international career as a popular music producer of 30 years, holds a doctorate of education and lectures in music production within the Bachelor of Music at the Queensland Conservatorium, Griffith University. Brendan's research engages with popular music production, popular music education and the relationship between technology and music production creative practice. More information is available at brendananthonymusic.com.

"Brendan Anthony has produced a visionary, forward-thinking book which is not only fascinating, entertaining, and highly readable, but also rigorously academic. Drawing from an amalgam of professional experience across the music industry and higher education divide, he has woven a detailed and erudite research project through the two. A quite brilliant achievement which I hope will be a landmark in how higher education and industry work for popular music production in the future."

Professor Lucy Green, *UCL Institute of Education, London, UK*

"In this exciting new book, Brendan Anthony presents a timely and detailed student-centred exploration of teaching and learning in popular music production, engaging producers, performers and professors as he weaves together expert pedagogical advice and profound insights from both sides of the glass."

Dr Gareth Dylan Smith, *Assistant Professor Music, Music Education, Boston University, USA*

"Brendan Anthony has made a significant contribution to the world of pedagogical music production practice with this volume of work that will be celebrated by educators internationally. This authoritative work has been written from both the perspective of an academic and a working practitioner. It provides an insightful and much needed view of what goes on behind the glass from the lens of an educator and highlights the key issues for academics and students working in the field. The field of music production continues to develop from both technical- and cultural-perspectives that explore both the creation of the technology, its use, and the approach of musicians and record producers and how they create work in the studio. This important text and much needed work will fill a gap in the literature in terms of how we frame popular music production methods from pedagogical and cultural perspectives."

Professor Andrew King, *Head of School of the Arts, University of Hull, UK*

Perspectives on Education in Audio and Music Production

Series Editors:
Russ Hepworth-Sawyer, *York St John University, UK*
Mark Marrington, *York St John University, UK*
Tim Canfer, *University Campus Barnsley, UK*
Carola Boehm, *School of Digital, Technologies and Arts, Staffordshire University, UK*
Brendan Anthony, *Queensland Conservatorium, Griffith University, Australia*

Perspectives on Education in Audio and Music Production (POEM), centralises discourse around teaching and learning within the fields of audio, music technology and music production, including creative music creation. The series collects both academic and experientially informed considerations of teaching and learning from a multitude of perspectives. POEM captures not only the views of the educator, but also the perspectives of the learner as well as the industry professional. Issues around historical and contemporary context, pedagogical issues, self-directed learning paradigms, employability and vocationality, professional education, and industry relations with education provide the foundationary themes for the series.

Music Production Cultures
Perspectives on Popular Music Pedagogy in Higher Education
Brendan Anthony

For more information about this series, please visit: www.routledge.com/Perspectives-on-Education-in-Audio-and-Music-Production/book-series/POEM

Music Production Cultures

Perspectives on Popular Music Pedagogy in Higher Education

Brendan Anthony

Routledge
Taylor & Francis Group
LONDON AND NEW YORK

Designed cover image: Brendan Anthony

First published 2023
by Routledge
4 Park Square, Milton Park, Abingdon, Oxon OX14 4RN

and by Routledge
605 Third Avenue, New York, NY 10158

Routledge is an imprint of the Taylor & Francis Group, an informa business

British Library Cataloguing-in-Publication Data
A catalogue record for this book is available from the British Library

Library of Congress Cataloging-in-Publication Data
Names: Anthony, Brendan, author.
Title: Music production cultures : perspectives on popular
music pedagogy in higher education / Brendan Anthony.
Description: Abingdon, Oxon ; New York : Routledge, 2023. |
Series: Perspectives on education in audio and music production |
Includes bibliographical references and index. |
Identifiers: LCCN 2022027861 (print) | LCCN 2022027862 (ebook) |
ISBN 9780367517649 (hardback) | ISBN 9780367517632 (paperback) |
ISBN 9781003055099 (ebook) Subjects: LCSH: Popular music—Instruction and study. |
Popular music—Production and direction. | Popular music—Writing and publishing. |
Education, Higher. Classification: LCC MT67 .A67 2023 (print) |
LCC MT67 (ebook) | DDC 782.42/13—dc23/eng/20221004
LC record available at https://lccn.loc.gov/2022027861
LC ebook record available at https://lccn.loc.gov/2022027862

ISBN: 978-0-367-51764-9 (hbk)
ISBN: 978-0-367-51763-2 (pbk)
ISBN: 978-1-003-05509-9 (ebk)

DOI: 10.4324/9781003055099

Typeset in Times New Roman
by codeMantra

Access the Support Material: www.routledge.com/9780367517632

Contents

Figures

Tables

Foreword

There was a time when record production was thought to be a skill acquired almost by osmosis. Those who believed they had a shot at being a professional music producer listened to a lot of music, played it some, wrote the occasional tune or two, hung around recording studios and nightclubs and eventually talked their way into an opportunity to work alongside a band. Lessons were learnt the hard way: by trial and error. In honest moments, successful producers often admitted that their careers owed a lot to the hand of fate. This outcome was caused by the fact that record makers who worked in popular styles, unlike their peers in jazz and classical, typically had no formal schooling.

For a long stretch of popular music history, from the 1970s until recent times, engineers and producers learnt their craft through the old-fashioned apprenticeship model – veterans took novices under their wings and brought them up through private, personalised tutelage. Those days are behind us. As Brendan Anthony's book can testify, the contents and the challenges of a comprehensive program in popular music production can be addressed in college or university.

A little history helps us to understand the challenges. The road to production starts on a path that leads elsewhere. There are four paths through which popular music producers typically pursue their careers. Each path originates with the acquisition of specialised skills and knowledge learnt in school or, in previous eras, in the trenches of music making. Those intent on a production career must parlay their expertise in one area for a chance to try their hand at a production job. If they achieve even moderate success, a second chance can follow and they're on their way.

One common path to a production career is taken by the arranger/composer-producer, best typified by Quincy Jones, who started as a bandleader in the 1960s before becoming one of the world's most renowned record makers. Formal music training in arranging or composition gives this producer the music/theoretic knowledge and the skill of coaxing musical pieces into a satisfying whole. The second path is the performer-producer. Many of today's producers begin their careers as studio musicians, touring artists or DJs. Expertise on piano or guitar or beat-making lets this person partner with a songwriter to make demos that lead to record deals and, often, the chance to produce. The third path is also common: the engineer-producer. This producer starts her career by learning audio engineering and signal flow. After building a discography with credits in engineering and mixing, she is often asked to take the producer's role and help a novice artist develop professional-level studio performance skills. The fourth path isn't well travelled today but was common in the 1950s and 1960s. This is the executive-producer. Record label executives and Artist & Repertoire (A& R) representatives once shepherded the making of popular records directly, in the studio. This changed in the late 1960s when artists began wresting creative control away from executives and putting it into the hands of those with greater musical or creative expertise.

The four critical knowledge bases – music theory, performance, audio engineering, music business – that can lead to a career in popular music production are increasingly seen as opportunities for higher education to draft curricula combining these disciplines under one umbrella. An inclusive production program spanning both the art and the craft of record making allows educators to focus on the spheres of knowledge crucial to every profession from medicine to politics: tools & methodology; the discourse community; the field.

Higher education music production programs give precedence to teaching record-making tools and the ways they are used. The swift advance of technology mandates that curricula must be frequently updated, and that instructors, out of necessity, must keep pace as scholars and practitioners lest they fall behind.

The influence of the discourse community can't be understated because popular music is, almost by definition, a reflection of its time. The rhythms supporting our popular dances, the latest local slang, the appetite for dissonance and the most cutting-edge sound design all pass through record makers' collective filter. Modern record producers introduce new musical ideas to the public by plucking certain ones from the cultural and technical incubators. Record makers are thus taste makers. Successful production students learn the importance of keeping pace with socio-cultural trends affecting the music of our times.

The field, of course, is music itself. Every young record maker must become a "student of the game" and have an encyclopaedic knowledge of the seminal records in his or her chosen musical subgenre. Students who know their history are better prepared for the startling speed of popular music's stylistic evolution. Production programs are challenged when designing courses around specific musical exemplars due to how quickly even the most groundbreaking records can become irrelevant.

Yet, viewed from the theoretical perspective of what producers must know and be skilled at performing, these obstacles are no different from those faced by schools of law or medicine. These challenges can be met and overcome; popular music production pedagogy is both doable and timely. This book takes an important step towards a long-awaited formal method of instruction in popular music production.

Susan Rogers, PhD, Berklee College of Music

Author's Preface

"Breaking the Glass"

MY STORY

I have always been fascinated with music; I used to play melodies on the piano by ear when I was five and I performed regularly as a vocalist and multi-instrumentalist throughout my school days. However, I will never forget the day my late father brought home a reel-to-reel tape machine, proclaiming that it did *sound-on-sound* recording. I was in high school at the time and I subsequently spent hours tinkering and recording with this machine before deciding to study Music Technology in 1985. University life further promoted my interest in the composition, performance and recording of popular music. I became absorbed in the music-making process and subsequently spent many nights in the university studios attempting to develop my craft. It was clear I had found my calling. I was obsessed with those who worked on the music that I loved and decided I wanted to be a music producer.

Following four years of study, and countless all-nighters in the university studios, I was lucky enough to land my dream job. I was to be the new assistant engineer at Rhinoceros Recordings (Rhino), one of Australia's most prestigious recording studios. It was the end of 1988 and after graduation I moved to Sydney.

EARLY 1989

As an assistant engineer, I worked around 100 hours per week (yes really) and got paid approximately $200 (AUD) per week. To be truthful, due to the calibre of people I was working with, I would have done it for free. The assistant engineer's responsibilities included setting up and packing away all the equipment, getting the lunches, making the tea and coffee and basically being there to do anything necessary for the betterment of the recording. I spent about three years at Rhino learning the practice of popular music production (PMP) whilst working with renowned producers and artists like Chris Thomas, Warne Livesey, INXS and Midnight Oil. During these years, I learnt all about sound engineering, mixing, musicality, performance, emotional transference, communication, psychology, songwriting, music programming, project management and the music industry. It also became clear to me during this traineeship that my four years' education at university had helped me learn a lot about sound and how to use technology, but I had learnt very little about the *practice* of PMP.

One of my first gigs at Rhino was working with producer Chris Lorde-Alge on Ian Moss' *Matchbook* album (Moss, 1989). Ian Moss is a renowned musician from iconic Australian rock band, Cold Chisel. However, "Chisel", as they are affectionately known here in Australia, are made up of two singers: rock legend and lead vocalist Jimmy Barnes and guitarist/vocalist Ian Moss ("Mossy"). This was Mossy's first solo album, and there was subsequent pressure regarding his status as a solo performer and singer. I recall a recording session where Chris was really pushing Mossy during the lead vocal performance for his song, "Mr. Rain".

CHRIS LORDE-ALGE: "Let's try that again, but this time really tear into that top note in the chorus and hold it for at least 3 to 4 beats".
IAN MOSS: "I'll give it a go mate but I'm not sure if I can do that."
CHRIS LORDE-ALGE: "Well let's just find out shall we?".

It was evident to me that even though Chris nurtured Mossy by complimenting him when he did well, he also really pushed him as a vocalist, challenged him to do things that were at the limit of his vocal ability and doggedly made him do lines repeatedly. I sat there and listened as the two went back and forth with discussions before more vocal takes ensued. After it was all over, I remember how good I thought the vocal performance was. I recall many social engagements or creative conflicts such as this during the making of that record and they all promoted the manifestation of a high-quality album. On reflection, I now know the five weeks I spent working on *Matchbook* were instrumental in the development of my understanding of PMP as a form of practice.

At Rhino I was like a socio-musical sponge. I studied how world-renowned producers worked. I listened as they pressed the talkback button, looked through the control room glass and communicated to the artists. I watched and listened as amazing performances were delivered and seminal albums produced. I was absorbed in the intense conflicts, outrageous parties and comedic banter that made up the social atmosphere of these sessions. Truly, it was a very special time in my life, one that I value to this day. It was here that I began to understand that PMP was influenced by three elements: (1) the music, (2) the technology (including instruments) we use to make the music and (3) the social, psychological and communicative engagements that take place during the music making. The skill of the producer was knowing how to utilise these three elements to achieve the correct balance of the artist's and producer's intentions in the recorded work.

One of the most influential producers I worked with was Chris Thomas. One night, after we had finished work on INXS's album *X* (Farriss et al., 1990), and were having a few drinks at a local King's Cross bar called Springfield's, Chris took the time to explain to me about the importance of balancing a producer's and artist's engagement: knowing when to let the musician do their thing and when to direct them. "You just have to use your instincts" he said, "you always need to have an opinion, and if the artist is floundering around get in there and instruct them. However, if they are in there making *gold*, no matter what you think, stay the hell out of their way!" Through my eventual agency as a producer, and over time, I came to realise that the combination of these two domains – producer and artist/ musician – was paramount to high-quality PMP taking place.

At that time in my life, I was also an aspiring artist and hadn't decided on a definitive career path. But it was made clear to me on many occasions that I had to eventually make a choice. The engineers who trained me at Rhino would often lay this out after sessions whilst we spent hours talking about how things should run and the role of the engineer/producer. I was explicitly told that producers discuss the music with the artist, engineers do not. Rather, engineers run the technology and they are like ghosts. However, inside me I also had this yearning to be a songwriter, artist and performer. I was frustrated in the engineering role because of my musical abilities. These silos of PMP (producer, engineer, artist) have traditionally been distinguishable, and my problem was having to choose which side of the recording studio glass (or which silo) I wanted to sit in. I eventually spent so much downtime in the studio recording other bands that production took over and, since that time, I've had a 30-year career as a music producer and mixer. I've never stopped writing music, but the idea of becoming an artist felt increasingly unobtainable as time went by and more production work came in.

MODERN PMP

If we fast forward to 2022, the evolution of the digital audio workstation (DAW) has often caused the producer's role to blur between songwriter, performer, producer and engineer, and I have engaged with all of these in modern music making scenarios. Artists can, and very often should, be encouraged to produce their own music, produce other peoples' music and engineer and mix music. The glass that once separated the producer from the artist is now metaphorically broken. The three elements (music, technology and social atmosphere) that I have listed above still constitute the practice of PMP, but the role of the music producer has many diverse and different connotations.

It was unfortunate that the university education I, and I assume many others, went through only focussed on the science of sound and the technology used in PMP. I mean, I get it: it is easy to talk about (and educate others on) what technology can do and how to use it in a creative way. But what I have learnt as a music producer is that the real magic in music creation happens when you consider *why* you would work, with technology or people, in certain ways for the sake of the music, and how you can connect with people during the process and through the development of the musical work. These are advanced processes and, as such, it is difficult to facilitate them in higher education (HE). This book is intended to aid educators in facilitating the teaching of these advanced concepts with the future music makers of our world. Moreover, I do not wish to limit the book to the ways we currently make music because this does not promote the longevity of the intended message. The educational framework outlined in this book promotes that music production should be learnt as a *holistic practice*. Therefore, the underlying themes of this book will have applicability in many different musical situations around the globe and into the future.

Finally, my background is that of a popular music producer who has been willingly enticed into the world of academia. Presently I am an academic who has completed rigorous research in the field of popular music production pedagogy (PMPP), but at heart I'm still a popular musician. Therefore, in the following chapters I will sometimes reflect on memories from my professional days and the lexicon of this book will, at times, shift towards vocabulary common in the popular music industry. I shall use words and phrases that connote popular music meaning and reflect the language of the profession (and so will many research participants). In some cases, these examples may be slightly colloquial, occasionally including mild profanity. This will not be prolific, and I trust this warning is sufficient to not cause any offense. Any video examples that contain profanity have also been identified to give prior warning (P). I do this to exemplify an important point: that the pedagogy of this field needs to align with the field itself. We must not be afraid to talk and communicate like popular musicians. As you read this book, you will see it is integral to the way music is made and to the pedagogy surrounding this. I trust I can find the most suitable balance of popular music lingo and academic voice, for it is my intention this book be representative of the popular music cultures we need to manifest within higher education institutions. So, with that said, Let's rock! \m/

Brendan Anthony working with students from the Queensland Conservatorium, Griffith University.

REFERENCES

Farriss, A., Farriss, J., Hutchence, M., & Thomas, C. (1990). *X* [CD]. [Recorded by INXS]. Sydney, Australia: Atlantic records.

Moss, I. (1989). *Matchbook* [CD] [Recorded by Ian Moss]. Sydney, Australia: Mushroom.

Acknowledgements

To my mother Teraze and sister Laine, I send my ever-eternal love and thanks. To my darling wife Kylie – thank you for being my rock, Mackas. There is no-one on this planet like you, you are an amazing woman and I love you. To my son Aussie – I hope you remember your young years coming into the studio, listening to the bands and their music, hanging, talking, laughing and discovering the wonder of music production. Embrace your passion and talent for music my boy, always follow your dreams and thank you for being my best mate. I love you. Dad.

INTERNATIONAL STUDENTS, EDUCATORS AND COLLABORATORS

I cannot give enough thanks to those who have taken part and helped me to collect the many perspectives I present in this book. I thank the students from my doctoral research and institutions from around the world and the legendary educators I spent hours talking to. A list of educators interviewed for this text include:

Prince Charles Alexander is a three-time Grammy winning, multi-platinum music producer, audio engineer and recording artist. He has worked with many artists including Mary J. Blige, Jennifer Lopez, Destiny's Child, Sting and Usher, to name a few. Prince Charles is currently a professor in the Professional Writing and Music Technology Division at Berklee College of Music.

Adam Patrick Bell is an associate professor in the Don Wright Faculty of Music at Western University in Canada. Adam is the author of *Dawn of the DAW* (Oxford, 2018), editor of *The Music Technology Cookbook* (Oxford, 2020) and editor of the journal the *Canadian Music Educator.*

Joe Bennett (Berklee College of Music, USA) is a musicologist, writer and researcher, specialising in songwriting and music copyright. As a forensic musicologist, he advises music lawyers, publishers, artists and songwriters on matters of plagiarism and musical similarity. Joe blogs about music technology, songwriting and copyright at joebennett.net

Samantha Bennett is a sound recordist, guitarist and professor of Music at the Australian National University. She is the author of *Modern Records,*

Maverick Methods, and co-editor of *Critical Approaches to the Production of Music and Sound*, and her co-authored book with Associate Professor Eliot Bates, *Gear: Cultures of Audio and Music Technologies*, is forthcoming from the MIT Press.

Gary Bromham is a songwriter, artist, producer and mix engineer. He has worked with many successful artists and songwriters including Bjork, George Michael, Sheryl Crow, U2, Bow Wow Wow, The Tubes, Robbie Neville, Eddie Money and The Fixx, to name but a few. He has just completed his PhD at Queen Mary University London where he is researching the role of retro aesthetics in music production.

Kaley Lane Eaton is a composer, singer-songwriter and producer from Seattle. Kaley's "disconcertingly lovely" compositions combine digital processes with ancient performance practices, questioning humanity's dependence on technology and the resulting exploitation of the planet. She holds a DMA in Composition from the University of Washington and is an associate professor at Cornish College of the Arts.

Andy Edelstein is a Berklee College of Music faculty member and former Music Production and Engineering assistant chair, as well as a Berklee Online course author. In addition, he works as an independent producer/engineer at Rapid Eye Music, a private recording facility in Belmont, MA, USA. His award-winning interactive exhibits can be seen at the American Jazz Museum in Kansas City, MO, USA.

Lachlan Goold is a recording engineer, producer, popular music educator, researcher and lecturer at the University of the Sunshine Coast. His research focusses on practice-based music production approaches, theoretical uses of space and cultural geography. Professionally, he is known as Australian music producer, Magoo, a two-time ARIA award winner.

Jan-Olof Gullö has a professional background as a musician (double bass / electric bass), record producer and television producer. Gullö now holds the position as professor and artistic and scientific representative in music production at the Academy of Folk Music, Jazz, and Music and Media Production, Royal College of Music in Stockholm, Sweden.

Russ Hepworth-Sawyer is a professional mastering engineer for MOTTOsound and part-time senior lecturer for York St John University where he has taught for nearly 15 years. Russ is also the founding editor of Routledge's Perspectives on Music Production series and is also the co-founder of the Innovation in Music series of conferences.

Kirsten Hermes is an interdisciplinary researcher, book author, senior lecturer, singer, violinist and audio-visual artist under the moniker Nyokee. She holds a PhD in sound perception from the University of Surrey (UK), and regularly publishes book chapters and academic papers. She recently wrote the book *Performing Electronic Music Live*.

Jay Hodgson is an associate professor of Popular Music Studies at Western University. He primarily teaches songwriting and record production.

Sören Johansson is a musician and teacher who studied at the Royal College of Music in Stockholm 1980–1985 and 2003–2011. Sören's expertise includes drum and percussion performance and music and video production. He was a member of Trance Dance (CBS/SONY) 1985–1990 and Peter Carlsson & Blå Grodorna 1993–2001. Sören has been an educator in media and music production at Dalarna University, Sweden, since 2001.

Alexis Anja Kallio is deputy Director (Research) at the Queensland Conservatorium, Griffith University. Her work examines the relations between music education, equity and justice. She is editor of *Difference and Division in Music Education* (2021) and co-editor of *The Politics of Diversity in Music Education* (2021) and *Music, Education, and Religion: Intersections and Entanglements* (2019).

Heather Kirby is a mastering engineer, educator and musician. She operates her studio "Dreamlands Mastering" in Prince Edward County (Ontario, CA), where she helps artists complete their creative vision and prepares their music to be unveiled to listeners. Heather teaches music production with RTA School of Media at X (Ryerson) University in Toronto.

Nyssim Lefford has worked in industry and academia, in artistic, engineering and scientific contexts. Her interdisciplinary research investigates the cognition of creative decision making. She holds degrees from Berklee College of Music and the Massachusetts Institute of Technology. She is a member of the faculty at Luleå University of Technology.

Jo Lord is an audio engineer, researcher and educator specialising in live sound and spatial audio. She has 15 years' experience as an industry practitioner across the live music, touring and event sectors. Jo has completed her PhD and her research targets 3D audio technologies for live music/record production. Jo also lectures in audio engineering at various institutions across England.

Mark Marrington is a senior lecturer in Music Production at York St John University, whose research interests include history of record production and classical music recording. His recently published book, *Recording the Classical Guitar* (2021), was nominated for the 2022 ARSC Award for Excellence in Historical Recorded Sound Research.

Phillip McIntyre is a professor at the University of Newcastle, Australia. He is a published scholar with ARC grants to his credit who researches creativity and creative industries. He is a songwriter, producer and audio engineer. As a manager he dealt with promoters, record companies and distribution labels. His own recordings are available on most streaming and subscription services.

Zack Moir is an associate professor of Music at Edinburgh Napier University, UK, chair of the ISME Popular Music Education SIG and a board member of APME. His edited books include *The Bloomsbury Handbook of Popular Music Education: Perspective and Practices* (2019) and *The Routledge Research Companion to Popular Music Education* (2017), amongst others.

Annelise Noronha is an Ontario-based producer, engineer, and composer and teacher. Annelise has been working professionally in Canada for 25 years. She has worked with artists including Jennifer Lopez, James Brown and Dragonette and assisting producers like Phil Ramone, Colin Linden and John Whynot. Annelise has also worked for Oscar winner Mychael Danna, and composer Lesley Barber. Annelise taught at the Harris Institute during 2019–2020, as well as many engineering and mixing workshops for clients such as the Canadian Women in Music organisation.

Amandine Pras is leading the Master of Arts in Music Production at the University of York (UK). A graduate of the Paris Conservatoire's Tonmeister program, she is conducting research partnerships to unlock the control room for women and non-binary producers, and other representatives of underrepresented social groups in the industry.

Toby Seay is a professor of Music Industry at Drexel University. His research interests include audio preservation practices, specialising in multi-track recordings and the study of sonic signatures within music production. Toby is the director of the Drexel Audio Archives. He has engineered multiple Gold- and Platinum-certified recordings, including eight Grammy winners.

Jarritt Ahmed Sheel is currently an assistant professor of music education at the Berklee College of Music. He is also a graduate of Teachers College, Columbia University's graduate program in Music and Music Education Department, and currently column editor for the MMEA Journal's culturally responsive column. Father, husband, son and friend.

Bryden Stillie is an associate professor in Music at Edinburgh Napier University, Scotland. His research includes developing new pedagogic approaches to teaching electronic and hybrid drum kit performance, exploring student transitions into Higher Education, Popular Music Education, and using technology to enhance the student learning experience.

Chris Stover is currently finishing a book for Routledge's Leading Change series, on reimagining music theory learning and teaching at the tertiary level. He is a composer and trombonist and senior lecturer of Music Studies and Research at Queensland Conservatorium, Griffith University.

Paul Thompson is a professional recording engineer who has worked in the music industry for more than 15 years. He is currently a reader at Leeds Beckett University and he is the author of *Creativity in the Recording Studio: Alternative Takes* (2019) and co-author of *Paul McCartney and His Creative Practice: The Beatles and Beyond* (2021) with Professor Phillip McIntyre.

Rob Toulson is a music producer and director of RT60 Ltd, who specialises in technology development for the music industries. He was previously professor of Creative Industries and Commercial Music at the University of

Westminster and Director of the Cultures of the Digital Economy Research Institute at Anglia Ruskin University in Cambridge, UK.

Sam Vallen is the lead guitarist and musical director of Australian progressive metal band Caligula's Horse, as well as a producer, songwriter and music educator. He holds a doctorate in music from the Queensland Conservatorium, Griffith University.

Mads Walther-Hansen is an associate professor in Music at Aalborg University, Denmark. He teaches courses in music production, music aesthetics and music entrepreneurship. His research is in sound analysis and listening with an emphasis on cognition and language in relation to recorded music.

Alan Williams holds a PhD in ethnomusicology and is a professor of Music at UMass Lowell where he also coordinates the Music Business program. He previously worked as a recording producer and engineer and writes and performs both with Birdsong At Morning and under his own name.

Simon Zagorski-Thomas is a professor at the University of West London and a composer, sound engineer and producer. He founded the 21st Century Music Practice Research Network and the Art of Record Production conference and association. He is the author of *Practical Musicology* (2022) and *Musicology of Record Production* (2014; winner of the 2015 IASPM Book Prize).

I sincerely wish to thank Sam Vallen for all his wordsmithery and advice whilst helping to finalise this book. I also want to thank Susan Rogers for her most considered Foreword, the series editors Russ Hepworth-Sawyer and Mark Marrington for their advice and Hannah Rowe from Routledge for forever being a positive influence on this book's creation.

All videos associated with this book are published with permission from Griffith University, Australia.

Introduction

Music production cultures and rationale

This book utilises a doctoral study to support the recommendations made throughout (Anthony, 2019). However, my doctoral work was inspired by both my engagement as an industry professional and my subsequent recruitment as a popular music production (PMP) educator. By working in both fields, I was located at a viewpoint that allowed me to identify the need for a scholarly investigation that targeted (1) the integral concepts and applications of modern PMP and (2) the ways that an educational experience may facilitate learning in these areas. Simply put, these two points outline the key aspects of the book you are about to read.

In the preface, I discussed how my experience as an assistant engineer informed my understanding of some key elements of PMP agency. These were: (1) the music, (2) the technology (including instruments) we use to make the music and (3) the social, psychological and communicative engagements that take place during the music making. Thompson (2019) emphasises that producers work through "numerous musical, technical, sociocultural and economic factors in their decision-making" (p. 3). Subsequently, the production process evolves and is reactive to all engagements within these areas. Zak (2001) writes:

> In short, tracks develop in an organic way, and distinctions among roles need not be definitive. Rather, collaboration fuels a catalytic responsive flow that can take many forms as ideas circulate among a project's participants and the project moves though its various stages.
>
> (p. 164)

The ability to let the production process evolve (as described above) requires high-level understandings of PMP creative practice and the enabling of intuitive and instinctive actions. Additionally, every musical work is influenced by the interactions of all the people involved within the greater socio-musical culture of the production. Therefore, no two production cultures are identical. The music, the technological affordances of the studio, the social and musical nuances of all the individuals involved and the cultural characteristics of the place that music making occurs (countries, cities, etc.) are unique in every production. The culmination of these

DOI: 10.4324/9781003055099-1

elements shapes the culture within which music is made and ultimately this influences the musical output. Through continued experience, professional producers learn that there is utility in designing and facilitating appropriate "music production cultures" that influence the musical work in beneficial ways. These attributes I accredit to a producer's tacit understanding of the greater *practice* of PMP.

As an educator, I began to realise that these concepts were not well represented in higher education (HE) frameworks. These can be complex scenarios for aspiring producers to grasp and this becomes even more challenging when one considers the environments that HE institutions provide for instruction. Therefore, through my doctoral investigation I developed a concept that suggested a student's engagement with diverse music production cultures enhances their ability in understanding, designing and engaging with the production process. The many educational approaches that I discuss in this book are all scaffolded upon my doctoral investigation where the utility of music production cultures was a significant finding.

WHAT ARE MUSIC PRODUCTION CULTURES?

To be clear, this book is not intended as a venture into the realms of cultural theory. However, I do believe that PMP is always undertaken within various music production cultures, and these cultures can be used as pedagogical tools. I will dial into these more explicitly presently, but for now it is suggested that music production cultures, like many cultures, consist of shared attributes. These are cultural considerations like the music; the song; the archetype of the producer and their process of envisioning; the social environment created by the people, places and spaces of the recording; the creative practice that is being undertaken; the technological influence on the creative practice; and the PMP skills being used within these cultures. As such, music production cultures encapsulate the totality of the PMP process, as shown in Figure I.1.

Music production cultures are not only a useful way of tabling a typology of PMP practice for students to view, they are also a suitable pedagogical tool that HE can utilise. Currently, I use the three layers in Figure I.1: skills, creative practice and cultural considerations as year level targets for students studying PMP. When students first come to university, I will develop a student's knowledge and ability in the skills. As they become modestly efficient at these, perhaps in the second year of their study, I introduce them to the many creative practice options they may encounter. When students have experienced these, I encourage them to engage in practice-based activities that consider the more advanced cultural considerations, shown in the outer layer, in their final years (see Anthony, 2021). There is no getting away from talking about all of the elements at any time, for example, first years may ask about envisioning. However, I do caution against overloading students with too much information early on as this can be detrimental to aspects of their learning (Student doctoral survey response, 2018), an observation I will explain more thoroughly when discussing student maturity and learning progressions later in the book.

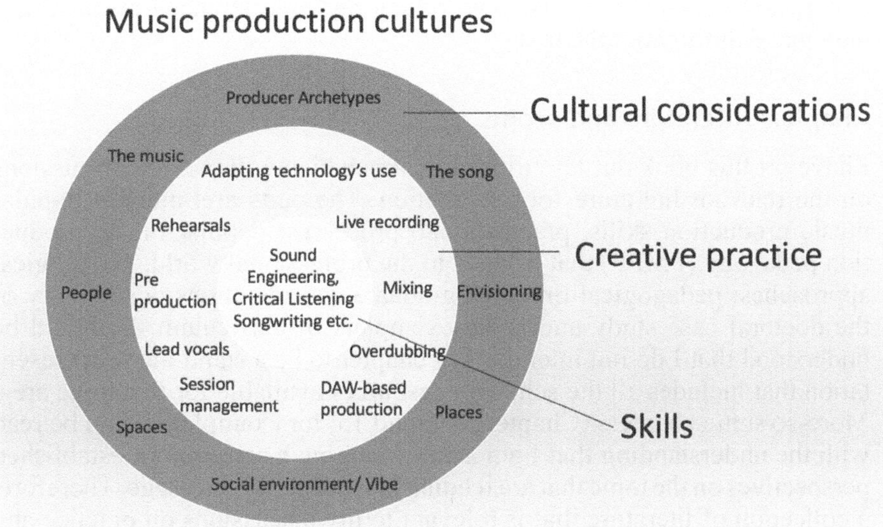

Figure I.1 Music production cultures (Adapted from Anthony, 2019).

I will also explicitly unpack the three layers in Figure I.1 and discuss the ways they can inform PMP educational approaches. The student cohort of my doctoral study (Bachelor of Popular Music students) overwhelmingly agreed that the utilisation of music production cultures as practice-based pedagogical frameworks was valuable in student learning (Anthony, 2019).

CONTEXT AND BACKGROUND

This book is firmly situated in the realms of recorded popular music which is recognised to incorporate lyrical and melodic content, be instrumental, includes acoustic and electronic genres and is separate from musical styles such as classical, jazz and musical theatre. However, should the characteristics of popular music change, the message within this book will continue to be a useful pedagogical resource into the future. This is because music production cultures occur in all forms of music making; they can be applied to film music, jazz, classical or musical theatre contexts, for example. Therefore, the overarching elements in this book would likely prove useful if nuanced and contextualised in other pedagogical settings.

This book is intended as a scholarly text which has been founded on rigorous research. This includes a doctorate of education (Anthony, 2019) that incorporated a year-long case study of two third-year PMP courses (see Appendix A for details before reading this book), survey data from 90 international students of all gender types, age ranges from 17 to over 36 years and where 95% studied between 2011 and 2021 (see Appendix C for survey questions) and qualitative interviews with 30 international educators. I will reference diverse literature and utilise data types that include quantitative findings, student and educator comments, video evidence of some of my classes and my own journal entries from the doctoral research. I trust the rigorous research I have undertaken prior to the writing of this book is understood and justifies any overarching and unreferenced statements I may make throughout the text.

Chapters, literature and scope

I have set this book out into different parts which will include discussions on the relevant literature for that section. The parts are: modern popular music production skills, practices and processes; popular music production pedagogy (PMPP) that bridges to the professional world; pedagogical approaches; pedagogical timeline, graduate considerations, a summary of the doctoral case study and some exemplars of curriculum. It should be understood that I do not intend every chapter to be a stand-alone representation that includes all the scholarly resources available for that topic area. More-so some chapters (Chapters 4, 5 and 15, for example) should be read with the understanding that I am acknowledging a selection of established perspectives on the topic that are feeding into the book's message. Therefore, a collection of literature that is relevant to my discussions on practice and pedagogy will be utilised and this is done to support the nuanced approach within this book.

RATIONALE: THE CURRENT STATE OF PLAY

I have been employed as a lecturer of music production at a university since 1998 (23 years' experience at the time of writing). In 2010, I was discussing with some of my students the great lengths to which some renowned popular music producers went to achieve authenticity in their music production. The song in question was Phil Collins' "In The Air Tonight" (Collins, 1980) and I was having a difficult time convincing my students that the legendary Phil Collins' drum sound (which utilised the SSL Listen Mic[1]) was, in fact, stumbled upon during the process of creative practice rather than being a planned event (Flans, 2005).

Hugh Padgham elaborates:

> One day, Phil was playing the drums and I had the reverse talkback on because he was speaking, and then he started playing the drums. The most unbelievable sound came out because of the heavy compressor. I said, 'My God, this is the most amazing sound! Steve, listen to this.'
>
> (as cited in Flans, 2005)

I distinctly remember the stunned look on some of my students' faces as they enquired as to how that kind of creative practice is learnt. I hesitated for a minute before suggesting that, following on from university, I, like many other professionals, developed an intuition for such dynamic and context-dependent innovations through experience. First, by working as an assistant engineer for a large studio with producers who engaged in the many creative practices of PMP. Second, my learning continued through my practice in both traditional and modern production landscapes. For the past 30 years I have operated as a freelance singer-songwriter, performer, programmer and music producer. I explained that historical normalities of studio production teams included an artist, a producer, an engineer and an assistant engineer. This is how the assistant engineer of the times learnt the fundamentals of PMP: by immersion in the process, and by watching the producer, engineer and musicians engage with the skills necessary for PMP. Subsequently, these aspiring professionals then honed their skills throughout their freelance careers.

It was obvious to my students and me that historical (apprentice-model) PMP education processes were very different from what they were experiencing within PMPP in higher education today. There are currently numerous HE programs[2] that offer training in music production principles, yet Frith and Zagorski-Thomas (2012) allude to the uncertain positioning of music production in academic frameworks: “What is less clear, though, is how the study of recording is being placed in academic departments and degree courses ... it was agreed recording usually fits rather awkwardly in its obvious home, departments of music” (p. 2). PMP is also a practical process, something that is essentially learnt by doing (Anthony, 2015). Therefore, it is important to contemplate how the ability to grasp the various musical, technical, social and psychological approaches of PMP equips the aspiring producer with the tools needed for practice. To a lay a platform to respond to these questions, I feel it is important to first identify and consider the transition that has occurred in HE sectors over the last 50 years, the programs students may be enrolled in that allow study into PMP and how this may influence any pedagogical approaches.

HIGHER EDUCATION INSTITUTIONS: HISTORY, TRANSITION AND LIMITATIONS

When I contemplated writing this section of the book, I was quite daunted by the diversity in HE approaches across the globe. This book does not seek to unpack these details explicitly, however, it is important that some context is provided here so later comments have a foundation to stand on. To be clear from the outset, I currently work for a conservatorium of music that follows what is commonly referred to as the *conservatoire model.* Often this term connotes an institution that focusses on performance and Western classical programs. However, the conservatorium with which I work operates within a larger university and has made attempts to become more progressive and nurture the future aspirations of all musicians.

I had a wonderful discussion with Simon Zagorski-Thomas who was knowledgeable in the area of institutional history.

> The way that education has changed in the past fifty years in the UK and to some extent in the States has been that, if you think about it as an industry that’s expanded, what has happened is, you’ve got these really well established ‘old school’ universities Oxford, Durham and Edinburgh they were all established in the sixteenth, seventeenth and eighteenth century. Then you’ve got ‘red brick’ universities in the nineteenth centuries [University of Birmingham, Bristol, Leeds for

> example]. Then in the sixties you had new comers … like Sussex, Stirling and Essex … you also get all these kind of new things coming in … they didn't compete against Oxford … they said, "right well what are the new subjects and what are the new approaches to subjects and what can we do that they don't do?". So, the ones that were successful were the ones that did artificial intelligence or electro acoustic music in the sixties. What happened around 1992 is we had this parallel system of vocational colleges called 'Polytechnics' and they were given university status. It's like the state colleges in the USA, they're mostly vocational. They're considered a lower strata of universities because they're about learning how to do a job, rather than learning a subject, like physics or the classics. At the older universities you don't get many degree courses that are named after a job … London College of Music was a conservatorium that was going for one hundred years, but it went bankrupt, so it joined one of these post 92 universities. So, for example, the University of West London is a post 92 university, Edinburgh Napier is a post 92 university.
>
> (Simon Zagorski-Thomas interview, April-May, 2021)

Simon went on to explain that the post 92 universities were exploring approaches seldom emphasised in academia: how to do something rather than how something is done. He also suggested that these styles of education require a different approach because values by which something is judged need to be clarified. These include subjective processes that qualitatively appraise "good" music and music production in a subjective space. Considerations such as these are not purely technical; they include aesthetic decisions.

> Within traditional musicology there is good technique and there is bad technique [for example, in composition] … [old school universities would propose that] "Once you've learned the 'right way', then you can go off afterwards and be creative". What we're trying to do [in post 92 universities] is to teach them to be creative, we're doing something that is very different from the old school universities.
>
> (Simon Zagorski-Thomas interview, April-May, 2021)

When I spoke to Jay Hodgson, he also mentioned the existence of a "university and college divide [in the United States and Canada], the idea is that college is professional". Jay went on to suggest that he studied at a trade school, and they are seen as a lesser form of education (Jay Hodgson interview, April-May 2021). It is unfortunate when hierarchy is placed on different HE institutions in this way, but the fact is that, when considering PMPP, the varying approaches of different institutions are challenging. Students not only need to learn how to *do* PMP, it is also integral for aspiring producers to use critical enquiry to consider *why* something should be done a certain way. This is primarily because PMP embraces many things that are creative, subjective and at times technically wrong: distorted and over compressed sounds, for example. Popular music producers most definitely need to learn the fundamental theories about sound and how to use technology correctly, but these are entry-level skills and not the important advanced PMP processes that institutions must facilitate. When considering advanced processes like envisioning, emotional transference, entrepreneurial workflows, psychology and communication it becomes evident there is no right or wrong way to *do* PMP, there are just different ways of getting an aesthetic result (Anthony, 2019).

Education in this field needs to nurture the student's development as people so they can be bold enough, whilst working in collaborative environments, to trust their subjective intuition and capabilities in the above areas. Subsequently, institutions that aspire to generate the music makers of the future need to immerse their students in the cultural realities of PMP, whilst they are making music to gather experiences of *practice*. Therefore, pedagogical approaches need to consider student-centred frameworks and encourage the socio-cultural foundations mutually common to popular music. This does not mean that PMPP cannot exist in all institutions; it most certainly can. However, the question remains whether equality in this type of education can be achieved within different institutions, primarily because courses purely focussed on the *right* and *wrong* way to do things, in my opinion, are better suited to the training of a *sound recordist* as opposed to a *music producer*.

In this book I will explicitly unpack much of what I have said above whilst considering the protocols and frameworks that institutions ask of their educators. I intend to explain my experiences and findings in this field and present some ideas to overcome problematic areas. To start with, it is pertinent to consider the differing programs that HE institutions offer where PMP can be studied.

PROGRAMS

Students from around the globe that study PMP do so from within a diverse array of undergraduate and postgraduate programs, many of which are seen in Table I.1.

Due to the different types of programs (listed in Table I.1) and their admission processes there is a very broad cross section of skills and affordances that students will bring to classrooms around the globe. If a program is performance-based, then applicants are usually required to audition and be proficient on an *instrument*, yet other programs do not require any type of audition for entry. If the program is specialised, like a bachelor of popular music, then it is advantageous if students are song-writers. Conversely, during my discussions with Adam Bell, he suggested that most of the students who take PMP courses at his institution are classical music majors and that some are not even music majors, studying science degrees instead (Adam Bell interview, April-May, 2021). Several of my interviews were with educators who also attested to the presence of institutional pressures to reach cap numbers and admit people who may not be completely suitable. The principle take-away from the above is that there is incredible diversity in the affordances of students enrolled in programs that facilitate PMPP

Table I.1 Programs from which students can study PMP

Undergraduate Program	Major/Minor (some examples)
Bachelor of science	Music technology, audio engineering and technology, sound engineering, live sound and engineering, music industry
Bachelor of music	Popular music, music technology, music production and engineering, sound production, music composition, performance, composition for film and video games, electronic music production and design, song writing, contemporary writing and production, classical music
Bachelor of arts	Music production, digital music, popular music, creative musicianship, music technology, creative business and music production, independent music production, media production
Bachelor of popular music Bachelor of creative industries Bachelor of sound and music production Bachelor of recording, mixing and production Bachelor of electronic music production	
Postgraduate programs	
Master of music	Music/record production, music performance, research
Master of science	Sound design
Master of arts	Audio Production, record production
Master of music production	Popular music

(International educator's and student's survey responses, May, 2021).

around the globe. This diversity not only needs to be considered in pedagogical designs, but it would also suggest that the needs of students may be considerably different across varying programs. Therefore, it is pertinent to understand the overarching reasons students may have for enrolling into university study that includes PMPP.

THE PURPOSE OF STUDYING POPULAR MUSIC PRODUCTION

I have worked at many *Open Days* for my institution and, in that time, I have done my best to encourage potential students (and their attentive parents) as to why they should come and study PMP at my university. In my opinion this can be undertaken, in most situations, on three levels: (1) my level – give the students an enriching experience and promote them being good people. (2) The parents' – get my child a job and secure their financial future. (3) The students' – teach me the cool stuff I don't know.

Give the students an enriching experience and promote them being good people

Popular music professionals do not enjoy working with someone who is difficult to be around. Samantha Bennett and I discussed this, and she eloquently described a truth about many graduates:

> The majority of it has absolutely nothing to do with music production and everything to do with how to be a decent person in this world … I've been teaching this subject for eighteen years … I've had cohort after cohort, three different institutions, two different countries … and there are dozens of students who can … boot up the studio … they can all record … they can all then turn that into something cohesive and relatively meaningful. But … it's only a few who have also got the reliability to show up, the ability to include other people [and] the ability to basically not be an arsehole!
>
> (Samantha Bennett interview, April-May, 2021)

I couldn't agree more with Samantha. By connecting students with some form of understanding of themselves and promoting the possibility of them developing into sociable people, educators nurture the potential of students becoming valued members of our society. Jarritt Ahmed Sheel and I discussed this further:

> What music teachers do … is they help people. Hopefully this is the goal: dependent learner to independent learner to being a good person while using the experience of learning about music and it's still used to emote and express through music.
>
> (Jarritt Ahmed Sheel interview, April-May, 2021)

The transition of learning that Jarritt speaks of (dependant-independent-good person) is transferable to many occupations so there is value in considering this as an important reason for studying at university. However, we are talking about popular music here, and I can hear the parents shouting loudly in my ear: "what about my child's employment prospects post university?"

Get my child a job and secure their financial future

In truth, students can become popular music professionals without going to university (Alexis Kallio, interview, April-May, 2021). So, what can university offer students that they can't get by living an entrepreneurial lifestyle? Rob Toulson suggested "you don't need to go to uni to be a music producer but you do need to go to uni probably to learn how to have a forty-year career as a music producer" (Rob Toulson interview, April-May,

2021). Longevity is important and due to the ever-evolving nature of the popular music industry (Burgess, 2008), our graduates need to be innovators and intuitive thinkers – people who find new ways to make a living. Andy Edelstein reflected:

> I had a student who had great success in the record business because he was in on the ring tones thing back twenty years ago, when people used to spend three dollars for a ten second ring tone. Think about it, the economics really worked but you had to be there at the right time. He was. He recognised the opportunity. He jumped in and made a business out of it, did really well, and then was able to support his production habit afterwards.
>
> (Andy Edelstein interview, April-May, 2021)

Therefore, in this book I will discuss how we can best facilitate education that promotes graduate industry integration and graduate industry innovation. I will support previous scholarly works that advocate for popular music education which includes increased domain immersion, an ability to be self-critical and edit one's work, to be genre-agnostic and creatively free and to emphasise the building of a portfolio of work (Bennett, 2015). I will also approach the bigger questions that ask what HE should look like, given the ubiquitous availability of information on the Internet. I will discuss how our courses need to leverage all possible resources (including Internet videos) but also focus on the things that these videos cannot provide. Rob Toulson articulates some of these things as: "bespoke, it's contextual, it's one to one, it's mentoring … it's teaching things like project management, research, critical reflection. All those things that you can't learn from YouTube or an online course. That's what universities have to offer" (Rob Toulson interview, April-May, 2021). However, it is doing our students a disservice to suggest that all of them will have a career as popular musicians that's simply not the case. We must get away from thinking that industry-readiness is the only reason to study PMP. Alexis Kallio discussed this:

> There is this narrative at university that you know what you want to be when you grow up and you need x, y and z and when you graduate you will be fully equipped for the real world to do that and get a job. None of that is necessarily true, through no fault of the student or the institution. It is undesirable to think that this is our only job as educators – to prepare students for employment. We don't know how the job market will change, or what skills might be required – especially if we consider advances in technology! There are jobs that we can't even imagine that young people will create themselves. There are many other institutions that prioritise preparing young people for the job market, and they should not be considered any lesser in status to a university. I have a real problem with this hierarchy of educational structures – they all offer different things, and we need them all.
>
> (Alexis Kallio, interview, April-May, 2021)

Zack Moir had some integral input on this subject:

> The biggest challenge we face is, to my mind, the employment-focussed, entrepreneurial, 'industry' mediated constraints of having to 'train' people to operate equipment. To me, so much of what happens in this world is mediated by 'industry standards', 'employability skills' and 'best practice'. Sure, some of these are important, but they are relatively mundane, and domesticating, and I feel they have far too much prominence in higher education, at the expense of genuine creative, exploratory, reflective and (most importantly) critical thinking.
>
> (Zack Moir questionnaire reply, April-May, 2021)

Jarritt Ahmed Sheel spoke to me about students who come to study music because it is a calling.

> You know, like, I could learn a vocation, but avocation is like a calling. Maybe somebody is … a doctor but they're like Yo-Yo Ma, on, you know the cello, but like you don't need to do it as a profession.… Like, how come you can't be a brilliant brain surgeon and also, you know, [playing the cello] like that's a calling but it doesn't have to be the person's way of making money … it could just be something they really do well.… We need to present all types of opportunities for students through music. It's not just about learning a skill to go and do it as a profession. [If we broaden the scope as to why people study in the first place] then we'll get more people doin' it lifelong. You know … 'cause it's just, it's dope man.
>
> (Jarritt Ahmed Sheel interview, April-May, 2021)

It is the PMP educator's duty to open our students' minds to all their future possibilities, and to remind them that these possibilities may include industry integration, industry innovation or a simple love for the practice of PMP.

Teach me the cool stuff I don't know

Many cohorts of music technology and popular music major students who study PMP with me arrive with similar attitudes. Half of them want me to teach them how to push and twiddle all the cool buttons, the other half want me to teach them how to make them sound like whoever is cool at the time: Ed Sheeran, John Mayer or Billie Eilish, for example. Students are also used to being fed information at school and many expect to be *taught* how to do things, but greater learning comes with an enabling of critical thinking. As Simon Zargorski suggested:

> It feels everywhere as if the commodification of education means that students and staff feel that the process of spoon-feeding students knowledge is what is required and actually that's not a way to get an

> education. The way of getting an education is to have someone who knows more than you, enable stuff to happen around you that allows you access to their knowledge.... I mean we stand in front of them and tell them stuff way too much. Because the students do feel that it's the 'hints and tips' type stuff that is their education, not the 'how to think' stuff.
>
> (Simon Zagorski-Thomas interview, April-May, 2021)

Alexis Kallio added to this discussion:

> I think university serves an important purpose: to be able to think critically, to be able to research, not in the form of formal research skills necessarily but in asking questions and looking deeper and having that capacity to imagine beyond what is, in a critical way. I think that is what university in general should be for.
>
> (Alexis Kallio, interview, April-May, 2021)

Over the last 21 years of facilitating PMPP I have found that once I have traversed some foundational learning with my students: how to work with sound, use audio technology and work within a DAW. We can then get to work on the things that really influence their music making: critical listening, critical thinking and reflection, considering the end listener and the subsequent function of their music, envisioning, increased domain acquisition, social skills, communication and psychology, generating performances, project management, being intuitive, etc. By encouraging students to engage in music production cultures that nurture the tacit development of these advanced concepts, students begin to see the bigger picture and their perceptions of that which is *cool* changes. I will go into these concepts in detail in the chapters to come.

PROFESSIONAL PRACTICE: BOTH SIDES OF THE GLASS

PMP has evolved tremendously over the last 100 years. Essential to the music producer's skillset is knowledge of an array of creative practice options spanning these different eras. PMP creative practice has also been interrelated with the development of technology. As technology evolves so too do the practices of music producers. For example, in the early days of recording a song was recorded as a live performance. With the development of analogue tape and multi-track recording systems, processes evolved that enabled producers to *drop-in*[3] during a performance and add many more overdubbed layers to a production. Finally, the DAW's random access editing capabilities have enabled more fragmented processes where producers can choose from a plethora of performance options. DAW software like Ableton Live, Logic Pro and Pro Tools also allows the producer to blur the lines between live performance and recording because input audio can be processed and manipulated in real time and recorded audio can be triggered live.

Each of these techniques is relevant today and they influence the recording outcome differently. Live recordings have been tried and tested for decades and are a way of designing a recording that generates something similar to seeing a band performing on stage. Conversely, sometimes the producer may use technology for the innovative processes it offers and this may deliver different artistic outcomes to traditional processes. Cher's single "Believe" (Higgins et al., 1998) used the technological development of *Auto-Tune* to dramatically alter her vocal sound and this became a signature characteristic of the song that many popular musicians now appropriate.

The development of technology has now altered the ways we can write and record songs. For example, Hip Hop and EDM producers utilise samples to create foundational beats or grooves within the DAW from which songs and productions are derived. However, technological use can dominate the creative practice of PMP and this can be to the detriment of musical and artistic outcomes. Moylan (2002) introduces his book by asking the reader to "think of a music recording as a piece of music, rather than a product of technological processes" (p. xxviii) and Brian Eno states, "as more and more options have become available, and equipment more and more complex, the temptation can arise to forget the possibilities of simplicity. It's a question of balancing external influences and technologies with your own instincts and beliefs" (as cited in Burgess, 1997, p. 85). These comments suggest that aspiring producers require the ability to reproduce genre or market-based sounds and concepts whilst having the foresight to use technology for its advantages while controlling its dominance on the creative process.

More recent studies confirm that PMP is the result of collaboration between experts in sonics, emotion and musicality who use practice-related knowledge to combine their resources and fold together a musical representation of a cognitive vision (Lefford, 2015). As a result, music production cannot be separated from the culture that surrounds it and the producer is akin to the captain of the ship, someone who is intent on "trying to find the delicate balance between the creative demands of the artist, the technical demands of the engineering staff, and the commercial demands of the record label" (Massey, 2000, p. xi). Massey's words are important because the end product can be produced with commonly used techniques and marketed in a multitude of ways in a multi-million-dollar industry (Hughes, Evans, Morrow, & Keith, 2016), or, alternatively, could be realised as something specific to the artist's musical intentions or self-expression and promoted to the public as such.

All the above is directly relevant to professional practice because PMP is the result of the interplay between all the people working on the music, and this holistic approach – considering production from *both sides of the glass* – therefore underpins much of this book. The phrase *both sides of the glass* is commonly used in the music industry to represent the glass in a traditional large-scale recording studio between the control room, where the music producer usually resides, and the recording room, where the artist performs. However, as I will discuss in detail later, the recording studio now connotes varying places of work: large scale, small scale, home and mobile studios (musicians can now completely produce a track on their laptop whilst at hotels or airports, etc.). While the metaphorical glass doesn't account for all such production environments, the term effectively situates the musical, social and technological processes of PMP as equals in this discussion. Within this book, I propose that to approach PMPP from the control room (traditional producer's), perspective alone only limits the authenticity of the pedagogical context to actual professional practice. The main reasoning for this is that I have seen firsthand the ease for which technological pedagogical approaches can be *dressed up* to look like instruction in PMP. Gary Bromham and I identified the danger in mantras that suggest technological knowledge is power. As educators, we must identify there is equal, if not greater, importance in the thousands of hours of practice some educators have done as industry professionals, or in practice-based research scenarios. These types of engagements enlighten the practitioner, and the message is loud and clear: the song (its function and intended audience), music, performances and the social relationships we develop with people as we make music are the centre piece of our practice. Technological application comes along for the ride – it is our vehicle and many of us love our vehicles – but there is immense value in practice-based scenarios which allow us to experience the socio-musical realities of PMP (Gary Bromham interview, April-May, 2021).

PMPP should bridge to professional practice to engage the aspiring popular musician with the necessary psycho-social skill development required for PMP (Anthony, 2015). It should also target the blurring of the lines between producer and artist in that those roles are now often combined into one (the bedroom producer), or shifted (the producer as co-writer). These are musical, technological and social practices that aspiring popular musicians need to experience to prepare for their futures, and, regardless of music type, they undoubtedly embody applications from both sides of the glass.

GLOBAL CONSIDERATIONS: INTEGRATING THESE CONCEPTS

As I stated in the preface, my experience of studying music technology in HE definitely did not prepare me for life as a popular music producer. My program covered many elements of sound, acoustics, electronics, recording principles, microphone theory and practice, critical listening, music theory and classical composition. It was post-graduation, working in the industry for 30 years, that my advanced education in PMP took place. In no way is that a negative reflection on the institution, educators and program that I studied under; I enjoyed learning all this *theoretical stuff* whilst I spent four years making popular music. My HE experiences were the result of the fact that currently there is no alignment between international approaches in modern PMPP. This book is proposed as the initial step towards developing that *foundation of practice*.

Within this book I advocate that PMPP should target the design of equitable and inclusive frameworks that enlighten all students to the greater *practice* of PMP (Anthony, 2019). This should be our overarching goal; effective education in this field does not culminate with students simply completing high-quality musical works. If students understand the *practice* of PMP, then, as they increase their abilities, they can tailor their skillset and creative approach to the musical endeavours that come their way. This interdisciplinary approach, then, provides the graduate with the potential for longevity as popular music practitioners. However, HE must also consider the explicit musical affordances (skills), music (Hip Hop, Folk and K-Pop, for example) and cultural characteristics of their students to develop student-centred approaches. Therefore, I am not going to suggest that the way I facilitate education to a group of Australian students at a conservatorium is the only way to do it. The virtues of various institutions around the world need to be considered and there are too many musical, cultural and social diversities in the places we live to do so. This is the reason I have interviewed 30 international PMP educators and collected survey responses from nearly 90 students from across the globe: to present varying *perspectives* on PMPP.

By writing this book, it is my intention to start a very important conversation for which many should have input. Gary Bromham suggested: "you know what needs to happen? You need to put everyone in a room together … and the egos need to be left outside the door" (Gary Bromham interview, April-May, 2021). So this is me inviting you into the room to have that first discussion, why don't we both leave our egos at the metaphorical door (as music producers are taught) and begin this conversation regarding a foundation of practice for PMPP at HE.

REFERENCES

Anthony, B. (2015). Creative conceptualisation: Nurturing creative practice through the popular music pedagogy of live recording production. *International Association for the Study of Popular Music, 5*(1), 139–156. DOI: 10.5429/2079-3871(2015)v5i1.9en

Anthony, B. (2019). *Perspectives on learning popular music production in higher education from "both sides of the glass"* (Doctorate of Education Thesis, Griffith University).

Anthony, B. (2021). The impact of diverse music production cultures on popular music pedagogy in higher education. In S. Holley, K. Reinhert & Z. Moir (Eds.), *Action-based approaches in popular music education* (pp. 107–115). Denver, CO: McLemore Ave Music & Flat Books.

Bennett, J. (2015). Creativities in popular songwriting curricula: Teaching or learning? In P. Burnard & E. Haddon (Eds.), *Activating diverse musical creativities: Teaching and learning in higher music education* (pp. 37–57). London: Bloomsbury.

Burgess, R. J. (1997). *The art of music production* (3rd ed.). London: Omnibus Press.

Burgess, R. J. (2008). Producer compensation: Challenges and options in the new music business. *Journal on the Art of Record Production,* (3). Retrieved from http://arpjournal.com/producer-compensation-challenges-and-options-in-the-new-music-business/

Collins, P. (1980). *In the air tonight [Recorded by Phil Collins] on Face Value [LP record]*. London: Virgin records.

Flans, R. (Producer). (2005). Classic tracks: Phil Collins' "In The Air Tonight". Retrieved from http://www.mixonline.com/news/profiles/classic-tracks-phil-collins-air-tonight/365521

Frith, S., & Zagorski-Thomas, S. (Eds.). (2012). *The art of record production: An introductory reader for a new academic field*. Burlington, VT: Ashgate.

Higgins, B., McLennen, S., Barry, P., Torch, S., Gray, M., Powell, T., & Cher. (1998). *Believe [Recorded by Cher] on Believe [CD]*. London: Warner Bros.

Hughes, D., Evans, M., Morrow, G., & Keith, S. (2016). *The new music industries: Disruption and discovery*. Cham: Springer.

Lefford, N. M. (2015). The sound of coordinated efforts: Music producers, boundary objects and trading zones. *Journal on the Art of Record Production, 10*. Retrieved from http://arpjournal.com/the-sound-of-coordinated-efforts-music-producers-boundary-objects-and-trading-zones/

Massey, H. (2000). *Behind the glass: Top record producers tell how they craft the hits*. Berkeley, CA: Miller Freeman Books.

Moylan, W. (2002). *The art of recording: Understanding and crafting the mix*. Boston, MA: Focal Press.

Thompson, P. (2019). *Creativity in the recording studio: Alternative takes*. Cham: Springer.

Zak, A. (2001). *The poetics of rock: Cutting tracks, making records*. Los Angeles, CA: University of California Press.

NOTES

1 The SSL Listen Mic was used for talkback between the control and recording room.

2 In this book a "program" is to infer an entire undergraduate degree (Bachelor of Music). A "course" is a subject a student undertakes as a part of their program requirements.

3 Drop in is a term used to describe the process where a segment of a live performance could be re-recorded so as to improve any flaws. This process enabled music producers to piece whole recordings together from many performances.

Part 1

Modern popular music production skills, practices and processes

This section of the book will outline a modern understanding of PMP so that later chapters can unpack how those elements can be facilitated within educational experiences.

DOI: 10.4324/9781003055099-2

1

Music production skills and creative practice

This book is not designed to advise on how to teach sound engineering, or how to introduce students to critical listening. These are foundational applications that I propose must happen prior to education on production. The main purpose of this book is to discuss how students can learn popular music *production* and, therefore, most of the pedagogical advice I want to outline will be centred around advanced learning outcomes (and, primarily, students in the latter years of their study). However, much of the advanced pedagogical materials that I want to discuss are facilitated as a result of foundational learning. Therefore, this chapter provides important discussions regarding the first two layers (skills and creative practices) encapsulated in Figure I.1.

LAYER 1: THE SKILLS OF POPULAR MUSIC PRODUCTION

The skills associated with PMP are discussed at great length throughout this book. As such, initial definitions of these will be useful to provide context for later discussions. *Critical listening* is a skill that involves listening on many levels and these include sonics, music and lyrical imaging (Draper, 2013). *Sound engineering* can best be described as the sculpting of recorded sounds with audio technology (Owsinski, 2014). *Musicality* is defined as a sensitivity to, and an in-depth knowledge of, music and music performance that includes pitch, rhythm, melody and music theory as it relates to popular music. The production of music often involves the social and musical interaction of people. Therefore, the music producer hones their *communication and psychological skills*. This is done (in one instance) to promote the capturing of performances (Howlett, 2009). *Emotional engagement* is a skill developed on many levels. The musician engages with emotion when generating performances; producers engage with emotion when critiquing performances and when designing sonic representations of songs using dynamics, equalisation and musical part interaction (Howlett, 2007). *Songwriting* is a skill that is prevalent throughout the production process from *pre-production* to *mixing* where producers target lyrics, melody,

DOI: 10.4324/9781003055099-3

rhythm, form, structure, harmony, arrangement and performance (McIntyre, 2001; J. Bennett, 2011). *Programming* is a skill where musicians use the computer's editing and compositional capabilities to arrange and produce sounds from within the DAW (Hugill, 2012).

Sound engineering

The holistic link between sound engineering, the science of sound and the setting up of recording environments has always been an important aspect for professionals to consider. To record music properly, one should understand how sound reacts within acoustic, electronic, digital and musical landscapes. It is also important for contemporary sound engineers to be comfortable with innovations in this field. The setting up of the recording studio environment (room acoustics, technological affordances) is essential to any practitioner. However, given this environment now occurs in diverse locations (homes, bedrooms, small buildings), these considerations need to be adaptable to the requirements of the work one intends to undertake.

The practice-based applications of various sound engineering techniques must be passed on to students. Following on from some educator-led exemplars of practice, students should engage in the applications of these skills individually and within groups to develop confidence in sound engineering. Concepts like microphone technique, pre-amplification and audio processing (equalisation, compression) and the use of effects (reverb and delay, for example) can be learnt from what I would classify as a *traditional* viewpoint. Following, it is always enjoyable for students to learn novel, exciting and perhaps theoretically *incorrect* ways of doing things (distorted vocals, for instance, that may be common in the music to which students are listening).

This is an ideal time for the educator to introduce the students to extant scholarly work on production and sound engineering. For example, educators can recommend texts like Schmidt Horning (2013), who discusses the importance of the recording engineer's tacit knowledge of microphone placement, and Izhaki (2008), who exemplifies practice-based concepts that link engineering applications, such as the use of echoes, reverbs and delays, to semiotic intentions like *life* and *distance*. Students often hunger for step-by-step processes of how to achieve results and there are many writers who use the final mix as a type of end goal of music production to illuminate the sound engineering steps involved during the recording process. Owsinski (2014) addresses sound engineering applications that include the fundamentals of microphone types, multi-tracking and the recording process – how to place microphones to record various instruments. Moylan (2002) explores the process of crafting a mix, dealing with various components including musicality, balance and spatiality. Finally, I find Gibson's (2008) book, which explores the concept of spatiality of a mix via a diagrammatical approach, very useful for entry-level pedagogical approaches on mixing.

Many students will be working in home studio applications and with the DAW, so instruction in sound engineering within these technological locations is also integral (Senior, 2014). Additional sound engineering concepts that may be suitable to varying student cohorts include information on suitable computer systems for DAW use, content that engages with surround sound applications, the online representation of audio (especially streaming), audio network devices, immersive audio, intelligent audio systems, audio object coding and the subsequent relevant requirements for audio in these instances.

A recent survey of international students found that 82% of students suggested that, as a part of their PMP courses, they engaged with *doing and learning* sound engineering and mixing at least 50% of the time, with 66% suggesting this happened most of the time. This is an encouraging result but not surprising because sound engineering, and the technological use that is associated with it, is relatively easy to facilitate in groups at HE, and this has been a consistent function of education in this field for some time.

Programming

The process of recording MIDI data that is then played through various hardware MIDI instruments was commonly known as *sequencing,* and this was a PMP skill that was prolifically utilised in the 1980s–1990s. With the advent of the DAW, musicians could expand this practice to include the use of extensive libraries of virtual instruments, audio samples and the production tools of the recording studio within music software such as Ableton Live, Pro Tools and Logic Pro. The editing power of the DAW currently allows programmers to sonically manipulate samples of audio. Audio can be tuned; time-compressed or expanded; edited; looped; equalised; filtered; reversed; and have various reverbs, delays, and modulation effects added. DAW's use music correction software applications like Auto-Tune and Elastic Audio that fine-tune the pitch and timing of performances. Another software application that is similar to Elastic Audio is Ableton's *Warp* function. Warping is incredibly useful because it can adjust audio in real time (upon input), is non-destructive, and makes audio manipulation a highly efficient and versatile compositional tool that is impacting on contemporary producer practices.

This book supports a definition of the term *programming* as an engagement with the compositional applications of both sequencing and the manipulation of, and composition with, audio samples. Given that this definition already constitutes some activities that could be classified as PMP, it seems logical that the agency of the programmer can often cross-over to that of a producer. The DAW's integration of random-access recording affords this even further by giving the programmer the technological capabilities to engage with producer-related activities such as recording acoustic parts (vocals, guitars, drums), generating performances with musicians and mixing/mastering their work. As a result, the popular musician/producer currently has the technology to compose, program, record and mix their music within the computer itself.

Programmers benefit from knowledge in musicality, songwriting, sound engineering, mixing, electronic instruments and synthesis. Some of these elements can be learnt in stand-alone courses (songwriting and sound engineering, for example), but there is also a lot to consider regarding the facilitation of programming pedagogical frameworks. This is because the social realities of programming situate it primarily as a solo practice, oftentimes in private locations. Elementary practices (DAW editing, for example) can be learnt in group contexts, but a programmer will regularly work alone. Similarly, processes like synthesis and the use of virtual and hardware technology can be learnt autonomously, following some elementary instruction, through trial and error, by accessing instructional YouTube videos (Justin, 2018), by reading texts similar to Puckette (2007) or reading weblogs similar to Bell's (2017a) "The pedagogy of Push: Assessing the Affordances of Ableton's 'Instrument'".

Programming has become a sought-after commodity because the process is omnipresent in electronic music genres. Seminal pop producers like Max Martin utilise it within their creative practice and some form of programming is evident in most popular music today (Collins & d'Escrivan, 2007). Programming technology has therefore revolutionised the music creation process (Frith & Horne, 1987) and given popular musicians the luxury of unlimited time to construct intricate compositions in home studios (Hugill, 2012). However, the aspiring music professional needs to consider the impact that programming has on music creation and eschew the practice of programming for matters of convenience alone.

A recent survey of international students found that 77% of students suggested that, as a part of their PMP courses, they engaged with *doing and learning* programming 50% of the time or less, with 50% suggesting this happened some of the time or never. Unfortunately, given the overwhelming amount of popular music that uses programming currently in the industry, this is a disappointing, if not surprising, statistic. It is no secret that many institutions have historically favoured instruction in acoustic-based music-making pedagogies as they are easier to facilitate in group contexts. Thompson and Stevenson's (2017) investigation into popular electronic musicians' perceptions and reflections of popular UK HE frameworks found that "the experiences of popular electronic musicians have been largely ignored in studies of popular music learning, which typically focus on rock-based musicians" (p. 204).

Critical listening, musicality and emotional engagement

Critical listening has been researched, discussed and taught from many diverse perspectives like *Timbre Solfege*[1] (Miskiewicz, 1992) and *Audio Signal Processing*[2] (Corey, 2010). I also find more practical approaches quite insightful. These works include Hugill's (2012) *Aural Awareness*[3], Walther-Hansen's (2020) discussion on the use of metaphors as descriptors of sound and Blake's (2009) perspective that producers are listeners and are "able to decode what happens in the recording and mixing studios in order to represent the eventual listening customer" (p. 36). It is vitally important that PMP educators target the diverse types of listening that popular music producers undertake. This includes the song's meaning, the sound box (the engineer's point of view); semiotic representations; the interplay between music and technology; and improvisational, melodic, harmonic and rhythmical relationships (Draper, 2013). In the book *Musical Environments*, Vella (2000) tables a guide for listening, composing and improvising that presents a multidisciplinary and sensory approach to making and perceiving music. Draper's (2013) and Vella's (2000) work confirms that PMP is much more than a technological process, it is a musical practice that often targets semiotic meaning. Therefore, listening purely for sound engineering characteristics like frequency response or compression is only one small part of the critical listening process. In this instance musicality is a constant point of reference for many authors.

Moylan (2002) writes "as technology advances … it seems that the artistic qualities of a performance have become more elusive" (p. xiii). Moylan specifically targets the relationship between technology and musical performance, and this is important because PMP communicates emotions to the listener by drawing "on the past experiences of the listener to shape their emotive reactions" (p. 62). People listen, stream and consume music (in one instance) because of the emotional response it may elicit in them. Zak (2001) also suggests "in the artificial atmosphere of the recording studio … the human emotion embodied by musical expression is easily lost" (p. 59). Popular musicians should engage with emotion whilst recording, but this has become even more difficult as technology has developed and afforded the producer computer-based creative practice options (Théberge, 1997). The implication of these authors' comments is that a fascination for technological advancement can affect the producer's engagement with musicality and emotion.

Within the concept of vocal recordings, Howlett (2007) suggests that the popular musician listens until "an emotional response is caused" (p. 2) and this signifies those "special moments" for which practitioners and popular musicians need to strive. Howlett (2009) goes further and states:

> there is an assumption made that, if it moves you, it can move other listeners. For the producer, trusting this emotional response is a quintessential function. The confidence to say, 'I like this one', is at the heart of a producer's role.
>
> (p. 17)

To be clear, musical listening (critiquing of performances, parts, harmonies, suitability of rhythmical elements and how various musical elements work together, etc.), and connecting to emotion when listening, would contrast with listening and only acknowledging the various instruments, sounds and words of the music. Knowing that popular music producers aspire towards the manifestation of an emotional reaction from the listener, it is certain that an engagement with musicality and emotion is integral in the decision-making processes underlying PMP.

Critical listening has been a consistent pedagogical target in many programs across the globe. These approaches include, but are not limited to, timbre solfege (frequency response) classes, musicological and semiotic listening discussions, peer feedback sessions where original material is played and critiqued and the constant engagement with listening whilst students apply their craft. The interconnection between critical listening, musicality and emotional transference is a useful model to integrate these three skills into pedagogical approaches and I will be discussing this later in the book.

A recent survey of international students found that 84% of students suggested that, as a part of their PMP courses, they engaged with *doing and learning* critical listening 50% of the time or more, with almost 25% saying they did it all of the time. Given the amount of research that has occurred in this area and the popularity of critical listening as a form of curriculum this is encouraging but not surprising. The same survey revealed that, when asked how often they engaged with musicality and excellence in performance, 57% of students suggested only some of the time or less, with 30% suggesting never. When asked how often the students engaged with emotional transference, results showed even less with 71% of students suggested only some of the time or less, and 36% suggesting never. Given the importance placed upon musicality and emotional transference in the making of popular music (as illuminated above), and knowing how important this quality is in the public's consumption of music, these are disappointing statistics.

Songwriting

Being adept at songwriting, or at least having knowledge of the elements and purpose of songwriting, is a definitive skill that is required within the practice of PMP. In traditional models of songwriting an artist would sit in a suitable/inspirational location perhaps with an instrument of some kind and compose a song. In this type of environment songwriters utilise knowledge of chord progressions, melody, rhythm, lyrical messaging and aesthetic/emotional intention or intended purpose (genre) for the song. However, the diversity that PMP creative practice now affords has opened new avenues for songwriting that contrast from traditional models.

Allan Moore's (2016) book *Song Means* analyses songwriting as a process that engages with *song, performance,* and *track* (recording). Moore defines *song* as a melodic, lyrical, harmonic, and rhythmic framework but then suggests the intricacies of *performance* (style, tempo, tone) and *track* (the recording of a particular performance) all have an influence on the way we perceive meaning from a song. The interrelationship between *song* and *sound recording* is ever present in modern songwriting applications. The DAW allows the *tracker producer* (see Auvinen, 2017) to engage in the songwriting process during the programming of a backing track for which a top-line writer then completes the melody and or lyrics. The DAW's influence on songwriting is unquestionable:

> DAWs are mediating structures – that is, they each have their own properties which influence how they may be used by the songwriter ... ultimately, how the DAW is used in songwriting will be determined by the place it holds in the process.
>
> (Marrington, 2017, p. 85)

Songwriting has become a diversified process in reaction to technological development. So much so that "popular song is arguably now more of a timbral and production language than ever before" and "the Internet affords different distributions of creativity across time and place, and changes what it means to create a song with others" (J. Bennett, 2018, p. 28). This has greatly affected the form that a song may take; J. Bennett suggests that "there are certainly characteristics of contemporary songs that are ... more common now than before DAWs and the Internet" and that songwriting as a form of agency "has verticality crossed from songwriter habit to listener preference" (2018, p. 26).

With the song being a primary influence on the way we make music, the *art* of songwriting is an integral element of PMP. J. Bennett (2015) suggests:

> [A]nyone immersed in the domain of popular songs will have developed an understanding of its conventions and constraints. The choice to break them according to artistic impulse is of course the student's own, but I would argue that a songwriter's skill is to make creative work within known constraints – not necessarily the constraints of the mainstream pop song, but of any genre in which the songwriter is working. (p. 41)

Songwriting involves a large amount of subjective decision making which can be applied within conventions or from a personalised perspective. Aspiring songwriters have both options to choose from and should learn to make informed decisions. The pedagogy of songwriting therefore, involves complex considerations, for this reason, songwriting is often taught in stand-alone courses (this is the case at my institution). However, this does not always occur, so it becomes integral to consider the role that PMPP can take in engaging students with the skill of songwriting within nuanced courses and programs.

A recent survey of international students found that 79% of students suggested that, as a part of their PMP courses, they engaged with *doing and learning* songwriting 50% of the time or less, with almost 61% suggesting this happened some of the time or never. This points to some problematic areas that may exist globally where PMPP is not accompanied by some form of suitable education or engagement with songwriting.

Communication and psychological skills

I would trust the front section of this book has highlighted my attestation towards the importance of the socio-musical environments in which music producers operate. Yet, in my experience it is often in these areas that HE students need to do the most work. Howlett (2009) is a practitioner-turned scholar who describes the producer using the metaphor of a *nexus*, suggesting that "the record producer acts as a means of connection between the artist, the technology and the commercial interest" (p. 1). In this landscape the producer needs to be able to communicate effectively with a diverse array of people: musicians, record company personnel, sound engineers, managers, radio interviewers, journalists, friends, family, children and even pets. If effective in these social interactions, the producer can steer the music production culture and this influences musical outcomes.

One of the producer's greatest challenges is to direct artistic approaches without controlling the musicians. To achieve this, producers use certain communication and psychological skills because "the way in which producers communicate their critiques and suggestions is perceived as more important than what is being communicated" (Pras and Guastavino, 2011 p. 84). During my industry training it became apparent that the psychological aspects of music production were often learnt through practice and by engaging with a vast array of musicians and personalities. Becoming adept at these skills often takes some time to develop because maturity and experience breed more confidence in these areas. Simon Zagorski-Thomas suggested:

> Most of the students are just not used to it, putting them in that context just makes them go "Oh I see, I get it. That's a person and I need to talk to them." They may be a bit clumsy to start with … but I think situated learning is a really important part of it – you put them in the context.
> (Simon Zagorski-Thomas interview, April-May, 2021)

In my initial communications with first-year students, I do suggest that some of the most important work we must do together is centred around their nuances as social beings and how they communicate. In truth however, the most influential work students can do in this area is done in their final years once students have matured but I target these skills from day one in a manner that reflects industry practice.

Students are always keen to pick up *tricks of the trade*, one of these that I pass on is to develop a rapport and trust with musicians so that when creative conflicts present themselves the artist respects the producer, values their opinion and this promotes high-quality communication practices. On the university campus I am always very social (not so much when I am at home, I am a recluse!). I am always complimenting students on their clothes; perhaps I will say something like "I dig your jacket, where did you get it?" I will ask what they ate for lunch, crack jokes, talk to them about their music or any upcoming gigs they may be attending. I am a sports fan so I always talk to them about sports if they are into that, if they are not into sports then we can joke about how different we are. By doing this, I develop a rapport with students, and get to gauge where they're at as a person and what kind of personality they have. I then use this information to nurture the vibe during tutorials or recording sessions. I will comment on their cool jacket and state that "I want one just like it!" I will use humour and perhaps tell them if they sing a tough line again "I will get them some pizza" (knowing they love pizza). These social interactions can distract artists from the pressures of performance in the studio and this helps them relax. Then I bring the students attention to how I have been utilising those everyday *social discussions* to develop relationships and nurture PMP creative practice. In this scenario students *get it* and see the value in this process.

Acting like a popular music producer when facilitating PMPP is something that I will talk quite regularly about in this book. I had a great chat to Andy Edelstein and he confirmed:

> Being a teacher is like being a producer in that there're many facets to the manner you might choose to go about exercising your craft, and also what you're best suited to do as an educator. Classes like ours are so different than classes I remember from college where you just sat there and listened to somebody speak.
>
> (Andy Edelstein interview, April-May 2021)

A recent survey of international students found a broad scope of responses to the request that they *"rate the amount [they] engaged with doing and learning communication and psychological skills during your PMP classes"*. 52% of students answered never/some of the time, 16% answered 50% of the time and 32% answered most/all of the time. Whilst it is encouraging to see a moderate to positive response here the majority of students are not engaging with these skills. There is need for improvement in this area to promote many of the student-centred graduate outcomes described in this book.

HARD AND SOFT SKILLS

The skills I have listed above can be considered under two headings – hard and soft skills. Hard skills are related to the requirements of doing a specific task where soft skills are "interpersonal and intrapersonal transferrable skills such as effective collaboration and communication" (Weston, 2020, p. 527). I discussed this concept with Adam Bell who has researched hard and soft skills within PMP. He compares the skills and practice of white water rafting to that of music production where there are *hard skills* – how to paddle could be likened to how to use music technology, add equalisation or compression or program a beat. Conversely, *soft skills* describe what we think and how we approach problem-solving and the nurturing of an experience.

In 2018 a colleague of mine conducted an online survey of 80 popular music graduates (graduating from 2005–2015). Following in-depth analysis, two themes emerged: "graduates were overwhelmingly engaged in what has recently become known as the *portfolio career*" (Bartleet et al. 2019, 2020); and that they valued the soft skills embedded in their learning as much as, if not more than, the hard skills" (Weston, 2020, p. 527).

Portfolio careers are characterised by entrepreneurial individuals who balance multiple concurrent roles and whose work is highly diversified. Their careers may be of high-risk compared to those in full-time employment and they face non-linear career progression (Bartleet et al. 2019, p. 285). In the era of the portfolio musician these soft skills are even more important given the uncertain future many of our graduates will encounter and the entrepreneurs they may need to become. Bell (2016) suggests that soft skills take considerably longer to learn than hard skills and that, subsequently, the "music educator should not be concerned with *commodities*, but instead *concepts* that are malleable to different learning environments amongst diverse populations" (p. 153). When designing PMP pedagogical frameworks, educators need to consider the learning progression of both hard and soft skills. These can be done separately, over varying time frames, and in more advanced circumstances where they are integrated.

LAYER 2: CREATIVE PRACTICE

Creative practice is a term used to describe the practical engagement that one undertakes during specific activities (Kerrigan, 2013). However, for the purpose of this book, creative practice will be used to describe the engagement of the people involved whilst performing, producing and recording music in some form of recording studio – all aspects of recorded music production. This includes approaches like improvising during a studio-based live recorded performance, mixing on an analogue desk, performing a guitar solo, programming beats inside the DAW or recording a lead vocal to many *playlists*[4] before *comping*[5] and applying Auto-Tune to the final version. Various creative practice applications can result in the manifestation of certain sounds or feels in the music they were used. For example, a producer who wants to achieve a musical work that resembles a 1960s record similar to "Light My Fire" by The Doors (Morrison et al., 1967) could use instruments and technology of that time (Fender Rhodes, Vox Continental, vintage microphones and an analogue tape machine). They may record the song live and inscribe it to vinyl for playback to best match the listening experience of the era. As Katz (2010) suggests, "repeatability may lead performers to mimic the recordings of others, or to study and learn from them" (p. 220). This is what the study of creative practice affords. However, Katz also states that "just as the technology shapes the activities of its users, their activities shape the technology" (p. 220). For example, in the 1970s when the turntable was recognised as a playback medium, the "New York DJs found that with a little tweaking it could be transformed into a performing instrument" (Katz, 2010, p. 220). As a result, *Vinyl scratching*[6] was born and identified as a new form of creative practice.

By promoting the role that technology plays in these scenarios, Katz supports the idea that "the value of any tool depends ultimately on its users" (Katz, 2010, p. 220). Given the vast number of technological advancements that have occurred in music recording technology in the last 50 years, the symbiotic relationship between technology and creative practice is an important topic for aspiring professionals to contemplate. The main problem with creative practice as a pedagogical target is that there are many ways to produce music and an education into all of these in three to five years is unobtainable. However, another approach, which has been prolific in the music technology domain, is to study the overarching influence that technology can have on creative practice.

Technology's influence on creative practice

Current research has explored the influence that technological development has had on creative practice, the subsequent options that become available to the producer because of this and how this affects musical outputs. Théberge writes "with the expansion of sonic technologies, the musician is able to engage with the micro-phenomena of musical sound itself, and such an engagement often forces a reassessment of the role of more traditional categories of musical practice" (1997, p. 186). An underpinning message from Théberge's work is that the music producer engages with technology, often governing its influence on creative practice so as to maintain intended musical intentions. Rap production provides a clear example of the producer using technology in innovative ways to aid the manifestation of a recording that targets the intended musical message. Rap production affords many related creative practices where

> [T]he equipment has to be altered to accommodate rap's use of low frequency sounds, mixing techniques revised to create the arrangements of and relationships between drums sounds... [and these are] strategies that revise and manipulate musical technologies so that they will articulate black cultural priorities.
>
> (Rose, 1994, p. 78)

These types of approaches are similar to the work of *Dub*[7] producers (Veal, 2007) who used audio technology in musical and creative ways. These techniques were often the result of trial-and-error applications and included adding delay to drum parts and physically kicking spring reverbs.

Frith and Horne (1987) describe the creative practices of music production prior to the technological revolution of the DAW as a social art form that utilised a technological process. However, with the advent of the DAW, a new dominant form of technology began to influence the way music is made. As a result, modern electronic music production utilises vastly different creative practice workflows than acoustic-based productions (D'errico, 2018). Nyssim Lefford discussed the evolving nature of creative practice:

> One could say that over the past 15 years, approximately, there has been a further, quieter revolution in production tool design, and as a result, what it means to "edit" and mix. Increasingly, production tools, processors, DAWs, etc. give the engineer/producer greater potential and power. The engineer/producer can increasingly generate more of the musical content that sits at the heart of any production. Furthermore, the lines between soft synth and signal processor, soft synth and DAW and sequencer and DAW are dissolving. All this only helps to further shift the production of musical content out of live rooms and away from recording.
>
> (Nyssim Lefford Questionnaire reply, May, 2021)

Nyssim's words signal clearly the responsibility on HE institutions to provide frameworks for students that are modern reflections of professional practice. As will be discussed later in this book, these practices currently reside in acoustic, electronic (DAW-based) and online realms (S. Bennett, 2018). Furthermore, it is our duty as educators to exemplify that practices are not constrained by technological applications. PMP manifests through technical and non-technical processes. Students must experience, work in and understand different domains of knowledge so that when the situation occurs in their professional practice, they can cherry pick from a broad scope of information.

When PMP occurs in online and via remote collaboration, it comes with a whole new reality of creative practice. This practice incorporates synchronous and asynchronous collaboration, project-based discussions over social media and other chat applications, file uploads via online storage platforms or software platforms like Pro Tools Cloud Collaboration that facilitates uploading content to multiple users automatically. Online and remote collaboration also changes the socio-cultural characteristics producers have become used to in face-to-face collaboration. As PMP diversifies further into the future it is time to well and truly recognise that the roles of producer, artist and engineer are often integrated, and departmentalised pedagogical approaches that separate these roles should be adapted. PMP educators must contemplate how their students can engage with learning the vast number of creative practices that are available to them, and perhaps more importantly, how they can learn which ones to select as options for the music with which they are working. It is the process of designing music production cultures which is integral for the aspiring producer to understand because this promotes the generation of authentic musical outputs as opposed to emulating something (Bell, 2017b).

LAYER 3: CULTURAL CONSIDERATIONS

The concepts of the outer layer of Figure I.1, cultural considerations, are more complex than the previous two and, as such, can be a challenge to facilitate in HE contexts. However, student engagement with these advanced concepts helps to stimulate music production cultures, thereby bridging the HE experience and the professional world. The next chapter of this book discusses the concepts of the outer layer of Figure I.1 whilst unpacking key elements of professional practice.

REFERENCES

Auvinen, T. (2017). A new breed of home studio producer? Agency and the idea 'Tracker' in contemporary home studio music production. *Journal on the Art of Record Production, 11*. Retrieved from https://www.arpjournal.com/asarpwp/a-new-breed-of-home-studio-producer-agency-and-the-idea-tracker-in-contemporary-home-studio-music-production/

Bartleet, B. L., Ballico, C., Bennett, D., Bridgstock, R., Draper, P., Tomlinson, V., & Harrison, S. (2019). Building sustainable portfolio careers in music:

Insights and implications for higher education. *Music Education Research, 21*(3), 282–294. doi:10.1080/14613808.2019.1598348.

Bartleet, B. L., Ballico, C., Bennett, D., Bridgstock, R., Draper, P., Tomlinson, V., & Harrison, S. (2020). Making music work: Sustainable portfolio Careers for Australian musicians. Retrieved from http:/makingmusicwork.com.au

Bell, A. P. (2016). Toward the current: Democratic music teaching with music technology. In L. DeLorenzo (Ed.), *Giving voice to democracy in music education: Diversity and social justice* (pp. 138–154). New York: Routledge.

Bell, A. P. (2017a). The pedagogy of push: Assessing the affordances of Ableton's "Instrument". Retrieved from http://www.adampatrickbell.com/blog/the-pedagogy-of-push-assessing-the-affordances-of-abletons-instrument

Bell, A. P. (2017b). Popular music pedagogy: 11 essential readings. Retrieved from http://www.adampatrickbell.com/blog/popular-music-pedagogy-11-essential-readings

Bennett, J. (2011). Collaborative songwriting: The ontology of negotiated creativity in popular music studio practice. *Journal on the Art of Record Production* (5). Retrieved from https://www.arpjournal.com/asarpwp/collaborative-songwriting-—the-ontology-of-negotiated-creativity-in-popular-music-studio-practice/

Bennett, J. (2015). Creativities in popular songwriting curricula: Teaching or learning? In P. Burnard & E. Haddon (Eds.), *Activating diverse musical creativities: Teaching and learning in higher music education* (pp. 37–57). London: Bloomsbury.

Bennett, J. (2018). Songwriting, digital audio workstations, and the Internet. In N. Donan (Ed.), *The Oxford handbook of creative process in music* (pp. 1–30). New York: Oxford.

Bennett, S. (2018). Intermixtuality: Case sudies in online music (re)production. In S. Bennett & E. Bates (Eds.), *Critical approaches to the production of music and sound* (pp. 231–252). New York: Bloomsbury.

Blake, A. (2009). Recording practices and the role of the producer. In N. Cook, E. Clarke, D. Leech-Wilkinson, & J. Rink (Eds.), *The Cambridge companion to recorded music* (pp. 36–53). Cambridge: Cambridge University Press.

Collins, N., & d'Escrivan, J. (2007). *The Cambridge companion to electronic music.* New York: Cambridge University Press.

Corey, J. (2010). *Audio production and critical listening. Technical ear training.* Burlington, MA: Focal Press.

D'errico, M. (2018). Technologies of play in hip hop and electronic dance production and performance. In S. Bennett & E. Bates (Eds.), *Critical approaches to the production of music and sound* (pp. 138–153). New York: Bloomsbury.

Draper, P. (2013). On critical listening, musicianship and the art of record production. *Journal on the Art of Record Production,* (8). Retrieved from https://www.arpjournal.com/asarpwp/on-critical-listening-musicianship-and-the-art-of-record-production/

Frith, S., & Horne, H. (1987). *Art into pop.* New York: Methuen.

Gibson, D. (2008). *The art of mixing: A visual guide to recording, engineering, and production.* Boston, MA: Thomson Course Technology.

Howlett, M. (2007). *Fixing the volatile studio: Studio vocal performance techniques.* Paper presented at the 3rd Art of Record Production Conference,

Queensland University of Technology, Brisbane, Australia. Retrieved from http://eprints.qut.edu.au/33275/

Howlett, M. (2009). *The record producer as nexus: Creative inspiration, technology and the recording industry* (Doctoral dissertation, University of Glamorgan). https://eprints.qut.edu.au/39093/2/c39093.pdf

Hugill, A. (2012). *The digital musician.* New York: Routledge.

Izhaki, R. (2008). *Mixing audio: Concepts, practices and tools.* Oxford: Focal Press.

Justin (Producer). (2018). Intro to synthesis part 1: Oscillators | reverb learn to play. [YouTubeVideo] retrieved from: https://www.youtube.com/watch?v=xyVFIPtOsm8

Katz, M. (2010). *Capturing sound: How technology has changed music.* Berkeley, CA: University of California Press.

Kerrigan, S. (2013). Accommodating creative documentary practice within a revised systems model of creativity. *Journal of Media Practice, 14*(2), 111–127. DOI: 10.1386/jmpr.14.2.111_1

Marrington, M. (2017). Composing with the digital audio workstation. In J. Williams & K. Williams (Eds.), *The singer songwriter handbook* (pp. 77–89). New York: Bloomsbury.

McIntyre, P. (2001). The domain of songwriting: Towards defining the term 'Song'. *Perfect Beat: The Pacific Journal of Research into Contemporary Music and Popular Culutre, 5*(3), 100–111. Retrieved from https://journals-equinoxpub-com.libraryproxy.griffith.edu.au/PB/issue/archive

Miskiewicz, A. (1992). Timbre Solfege: A course in technical listening for sound engineers. *Journal of the Audio Engineering Society, 40*(7/8), 621–625. Retrieved from http://www.aes.org/e-lib/browse.cfm?elib=7041

Moore, A. (2016). *Song means: Analysing and interpreting recorded popular song.* Burlington, VT: Ashgate.

Morrison, J., Krieger, J., Densmore, J., & Manzarek, R. (1967). Light my fire [Recorded by The Doors] on *The Doors* [LP record]. Los Angeles, CA: Elektra.

Moylan, W. (2002). *The art of recording: Understanding and crafting the mix.* Boston, MA: Focal Press.

Owsinski, B. (2014). *The recording engineer's handbook* (Vol. 3). Boston, MA: Cengage Learning.

Pras, A., & Guastavino, C. (2011). The role of music producers and sound engineers in the current recording context, as perceived by young professionals. *Musicae Scientiae, 15*(1), 73–95. doi:10.1177/1029864910393407

Puckette, M. (2007). *The theory and technique of electronic music.* Singapore: World scientific.

Rose, T. (1994). *Black noise: Rap music and black culture in contemporary America.* Hanover, NH: University Press of New England.Schmidt Horning, S. (2013). *Chasing sound: Technology, culture, and the art of studio recording from Edison to the LP.* Baltimore, MD: The Johns Hopkins University Press.

Senior, M. (2014). *Recording secrets for the small studio.* Burlington, MA: Focal press

Théberge, P. (1997). *Any sound you can imagine: Making music/consuming technology.* Hanover, NH: Wesleyan University Press.

Thompson, P., & Stevenson, A. (2017). Missing a beat: Exploring experiences, perceptions and reflections of popular electronic musicians in UK higher education institutions. In G. D. Smith, Z. Moir, M. Brennan, S. Rambarran, & P. Kirkman (Eds.), *The Routledge research companion to popular music education* (pp. 203–216). New York: Routledge.

Veal, M. E. (2007). *Dub: Songscapes and shattered songs in Jamaican reggae.* Middletown, CT: Wesleyan University Press.

Vella, R. (2000). *Musical environments: A manual for listening, improvising and composing*. Sydney: Currency Press.

Walther-Hansen, M. (2020). *Making sense of recordings: How cognitive processing of recorded sound works.* New York: Oxford University Press.

Weston, D. (2020). The value of 'Soft Skills' in popular music education in nurturing musical livelihoods. *Music Education Research, 22*(5), 527–540. DOI: 10.1080/14613808.2020.1841132

Zak, A. (2001). *The poetics of rock: Cutting tracks, making records.* Los Angeles: University of California Press.

NOTES

1 *Timbre Solfege* is the recognition of frequencies in sounds as related to tonal characteristics.

2 *Audio Signal Processing* involves critical listening that more broadly includes equalisation, spatiality, dynamics, reverb, distortion and delay.

3 *Aural Awareness* is a summary of the listening experience of the practitioner.

4 Within Pro Tools software, a *playlist* is a number of sub tracks that contain different performance versions of a specified track.

5 *Comping* is the process of selecting and combining various performance takes onto one audio track.

6 Vinyl scratching is a technique where musicians move a record back and forth with their hand to produce percussive sounds or samples.

7 *Dub* is a subgenre of electronic music that grew from reggae in the late 1960s.

2

Popular music production processes (Part A). Cognition, popular music production agency and socio-musical concepts

Understanding PMPP is contingent on a thorough grasp of the tools and mindset of modern producers. The next two chapters will outline the many considerations, concepts and processes that are associated with the production of popular music. I will do this in what may seem to be a sequential order – cognition, planning and manifestation. However, this is for written purposes only; the actual practice of production may be vastly different, and these elements may occur in any sequence, or even concurrently. Many of the professional applications I will discuss are represented in the outer layer of Figure I.1 and will inform the discussions of pedagogical application in the following chapters. This chapter and Chapter 3 are focussed on defining potential curriculum targets.

COGNITION, VISION AND THE AURAL LIBRARY: PLANNING THROUGH TO MANIFESTATION

PMP involves numerous types of cognitive work. As producers work to achieve favourable results, production goals direct metacognitive processes that inform decisions (Anthony & Lefford, 2019). Howlett (2007) suggests that "[a] production is a realisation of a creative concept" (p. 2) and the act of conceptualising is commonly referred to as a *vision* (Zak, 2001). Mixerman (2012) also states that: "vision ... is your ability to picture the final product before recording a single note.... If you don't have a clear concept of the project at hand, then you have no business producing it. Period" (p. 122). The key understanding here is the utility of cognitive approaches: when, how, why and to what purpose do producers use cognition?

The producer's vision

I recently undertook a case study investigation into the producer's vision (Anthony, 2017) and have developed the findings with my colleague Nyssim Lefford (see Anthony & Lefford, 2019). First, we found conclusive evidence that the producer participants (whom were experienced in both traditional and modern forms of PMP) all engaged with envisioning

DOI: 10.4324/9781003055099-4

regularly while they worked. Second, we found that to envision, or formulate a mental representation of what a production could be, requires many types of information including memories, formal knowledge, experience, etc. Since everyone's experience, memories and knowledge is somewhat unique, envisioning is an individualised process, something that is unique to us all. Envisioning had varied utility depending on the stage of production it was being used: planning, pre-production, tracking, mixing, etc. Regardless of its situation, it was clearly used throughout the entire production process. Another influence on the types of visions producers experienced were genre, past or current recordings, the song and technological affordances. The underpinning message that came from this study is that visions can be adapted and evolve depending on the progression of the music making. It also became evident that the envisioning process was important to the producer, and also, if they articulated their vision then all involved in the production process could have a clear idea of the intended outcome. Another theme that was generated as a part of this study was the influence of each producer's unique aural library.

The aural library

"Aural library" is a term appropriated from Schippers (2007). This term represents the musical, sonic and PMP memory that musicians have developed over time, from their engagement and interaction with music in all its forms – performance, listening and production. For a mature musician it is suggested that this may equate to 20,000 to 50,000 hours of listening (Draper, 2013) and, as such, the library that musicians can use for PMP influence becomes expansive. A popular musician's aural library informs their music production domain knowledge on which creative approaches rely (Thompson, 2019).

Planning through to manifestation

When starting a project, producers *connect* to the song's meaning or function to inform the development of a vision, and they utilise their aural library to shape this vision. Once this is achieved, producers switch their approach to planning the recording process so as to manifest the vision in recorded form. Again, I will reiterate that a vision is fluid and may change as the production unfolds. However, to plan the manifestation stage, producers use their knowledge of creative practice and socio-musical environments to facilitate the specific music production cultures within which they want to work. These cultures will be different for every production because every producer has an exclusive set of mental resources that informs their work and their envisioning (Anthony & Lefford, 2019). This knowledge comes from experience, and by becoming a music enthusiast – one who studies the ways that many producers have made records prior.

> Very rarely do I listen to a piece of music simply for pleasure these days.... I'm constantly deconstructing everything I hear, how that

> sound was created or why those instruments were used, or how the hell they managed to get the vocal to sound like that!!!.
>
> (Anthony Lycenko, as cited in Anthony, 2017)

Vision has great utility in the transition from planning to manifestation. In my experience, this is one of the hardest elements of PMP for students to master. The reason for this is because it requires the producer to have confidence in their own subjectivity and trust in their ability to make decisions and act no matter what creative obstacle may appear. This confidence comes with experience; it is developed over time and this requires specific consideration in pedagogical designs. Gary Bromham suggested that when he interviewed professional practitioners for his PhD, many articulated that they liked to utilise trial and error and emphasised finding an emotive response (or a *vibe*) from workflows.

> But you can't teach a student that – that's taste. Taste is something you develop, and everyone's taste is different. When you interrogate [professionals] further, you say, "yeah, but you must have tacit knowledge. You must have something there which you lean on?" and they go, "yeah, of course we do. That's a given" and we're going, "yeah, but hang on a minute". The given bit, we need to teach this part, but we also need to communicate the fact that there's an element of unpredictability, excitement and creativity. So, it's teaching the marriage of those things.
>
> (Gary Bromham interview, April-May, 2021)

The question remains how best to facilitate the development of the student's tacit knowledge base and their *taste* to inform the planning through to manifestation stages. There are frameworks that I have developed to target this (how to produce music), and this can be explained to students to help pave the way towards professional practice and the acquisition of tacit knowledge.

HOW TO PRODUCE MUSIC

The following concepts are typologies of my own creative practice. They are things I intuitively work through whilst producing music. They are methodologies I witnessed *in action* during my traineeship and put into practice throughout 30 years as a music producer. They are also useful pedagogical tools that I advocate to students as the three pillars of producing music.

Surveying of the musical landscape

The music producer first surveys the PMP landscape prior to the making of music. This important part of the production process is informed by any musical, lyrical, technological, cultural, commercial or expressive considerations. Surveying can occur cognitively. It can be part of envisioning, or it can happen in the form of research into a type of music, a cultural influence or a particular artist – that which has been done in this area before. This locates creative practice within some form of conventions (a genre, a style or an expressive intent) and sets some form of constraints on practice. This positions the producer's aspirations within boundaries that are tailored towards the end listener. However, the producer is not completely constrained since there are aesthetic choices the producer makes. Mixerman (2012) writes:

> The genre will dictate which conventions you adhere to, but music is far too complex to stay rigidly within stereotypes of a genre. Besides, how boring is that? Your goal is to create something stunning and new, but also recognizable. You must operate within and outside of convention simultaneously.
>
> (p. 123)

Considering the musical landscape in this way prepares the producer for the manifestation stages of PMP.

Designing diverse music production cultures: Making music in different ways

Following surveying activities producers use tacit knowledge to consider the music production culture within which they want to work. They will consider the type of studio, the technology and how long they will need to complete the project. Mixerman (2012) suggests:

> The recording process itself can and will carry great influence over the final product. Thirty days of recording time will yield a considerably different record from 10 days of recording the same band. The shorter time span with limited overdubbing gives you less time for polish. This will produce a far rawer product than that of a highly textured and painstakingly overdubbed product.
>
> (p. 124)

It is here that an expansive knowledge of PMP creative practice is beneficial. Producers will consider if the song should be an acoustic, electronic or hybrid[1] recording. This will inform technological choices, the necessary instruments and the people needed to play them. Explicit consideration must be given to the people required and their affordances (such as skillset and personality). Personalities should be emphasised in this equation as the social atmosphere of a session is an integral and influential variable on the musical output. The scope of possibilities here is expansive and, as such, a producer's vision should be decisive and inform a clear conception of how the music-making process *should* progress.

An example of a designed music production culture was the Foo Fighters Album *Wasting Light* (Foo Fighters, 2011). Interestingly, the Foo Fighters are credited as co-producers on this recording alongside Butch Vig, who was brought in because he was a known and trusted figure. The album was recorded in Dave Grohl's garage and was entirely recorded onto analogue tape with no computers involved in the process. Moreover, the band invited various session musicians to drop in and perform (Shepherd, 2011). Most importantly, the recording, because it took place at Grohl's house, enabled the band's families to visit. For example, kids could come in to the control room to say "Hi" to dad, and the vibe was nurtured (Moll, 2011). The result is a wonderful sounding record. In my opinion this is due to two factors. First, it had great mid-tempo, hooky rock songs. Second, the album sounded real and the performances had a raw energy to them, but the album sounded no less polished. These things come from working in environments that nurture such results.

Explicit application of popular music production skills

Once surveying activities have been undertaken and desired music production cultures put into place, skill application can then be explicitly informed. Practice becomes *dialled in* to the specific requirements of the music, the genre and any aspirations the artist may have regarding the song and the intended audience. For example, Rap music practitioners revised the use of the digital samplers to include *looped rhythmic lines* within their developing vocabulary. As Rose (1994) writes:

> Rap's primary focus is sonic, and the distinctive systematic use of rhythm and sound, especially the use of repetition and musical breaks, are part of a rich history of New World black traditions and practices.... Rap music centers [sic] on the quality and nature of rhythm and sound, the lowest, "fattest beats" being the most significant and emotionally charged.
>
> (pp. 64–65)

By working within an explicit music production culture, Rap producers are located in a domain of practice. Their programming, engineering and performative agency is applied from an informed perspective. However, even though the characteristics of Rap music can be targeted, innovations may evolve from these applications but the producer's work, whether appropriated or innovative, is influenced by the music production culture they are working within.

PRODUCER ARCHETYPES: MAKING MUSIC IN DIFFERENT WAYS

Before describing producer archetypes, a caveat needs to be introduced for this section of the book. Namely, in the next few paragraphs it is my intention to simplify some complex understandings of professional practice, in particular, to describe similarities in roles and approaches that are found among many producers. To do this I will utilise categorisations and typologies. I have also applied names to various categories, and I am aware of the connotations associated with the names we use. As such, I ask that this section of the book be read for what it is: typologies and theoretical propositions (or maps) that represent summaries of practice. I have found such simplified terms to be useful as pedagogical tools for PMP because, they do not exemplify, and therefore necessitate a familiarity with, the complex realities of the professional world. Also, this section in no way wishes to engage in the discourses of *auteur theory* or *craft and art perspectives* that are ever-present in the literature on the arts more generally. By suggesting that the below archetypes are viewed as typologies, I trust the use of associate terms will be deemed as suitable and useful. I do give particular thanks to Nyssim Lefford, Phillip McIntyre and Paul Thompson for their input on this and the creativity section in the book to follow.

Producers are inclined to make certain decisions and take certain actions over others, they value certain ways of working. Burgess (2013) discusses many typologies with which producers may identify and these include: auteur, facilitative, collaborative, enablative and consultative. Burgess' work describes the ways the typologies influence a producer's agency: how they will go about their work. Whilst working on some research with my colleague Nyssim Lefford on the producer's vision (see Anthony and Lefford, 2019) we similarly investigated the impact of producer *Archetypes* (Moorefield, 2005; Hepworth-Sawyer & Golding, 2010), but in this research, we considered how the adoption of an archetype could be a reaction to the production needs of the musical work. We then considered how the act of envisioning could be influenced by archetype, and how an archetype may influence the production process from conception through to manifestation. Two key differentiators among archetypes are particularly relevant for the producer's vision: first, how much value is placed on individual subjective preferences versus following musical conventions and norms; and second, how producers understand the producer's role and responsibilities (Anthony & Lefford, 2019).

Variation in PMP creative practices can be illustrated by three producer archetypes: the auteur, the innovator and the craftsperson. The first archetype is the auteur. This archetype is defined through their product representing their individual taste and subjective point of view. An example of a contemporary, highly technical auteur is Danger Mouse who is known for pushing the expressive potential of studio technology (Rambarran, 2015). The second archetype is the innovator. Archetypical innovators are interested in combining conventions from various types of music. Through this creative approach, *genre blurring* is becoming ubiquitous in the industry.

In the production of Amy Winehouse's "*Back to Black*" (Ronson & Remi, 2006), Marc Ronson and Salaam Remi fit this archetype. They carefully constructed vintage-sounding elements of 1940s jazz and 1960s Motown (Suddath, 2011), pulling from an existing field knowledge. At the same time, the material also deals with contemporary themes untouchable by 40s Jazz or 60s R&B, resulting in a very innovative album (Anthony & Lefford, 2019).

The third archetype is the craftsperson. Covach (2006) suggests a craftsperson "privileges repeatable structure; songs [that] are written according to patterns that are in common use" (p. 39). Craftspeople deliver what Marrington (2016) describes as "the relevant genre conventions (or codes)" (p. 205). The craftsperson aspires towards a consistent commercial "product", with a sound that attracts a certain audience. Craftspeople are strategic (Anthony & Lefford, 2019). An example of this archetype is Max Martin. To begin with, Martin was an innovator. As his reputation and signature sound developed, some of his methods became formulaic, and are now considered *normative* by imitators (Prime loops, 2012).

To be clear (and as will be discussed in the creativity section of this book), popular music producing operates under constraints, from the norms and conventions of style to a producer's own creative vocabulary. Such constraints impact all producers, regardless of archetype. At the same time, a producer's agency is not tied to one archetype. Ways of working are fluid, dependent on the musical context for which one is working. Burgess (2013) similarly suggests that "Some producers are bifunctional or polyfunctional and can operate in more than one mode" (p. 19). These archetypes provide guidelines from which producers can formulate their vision, design a PMP strategy and manifest their vision as a novel or familiar musical work. The archetypes are situated between concept goals that are biased towards intrinsic or extrinsic considerations and place hierarchal value on subjectivity, novelty and recognisability in different ways, as illustrated in Table 2.1.

McIntyre (2012) discusses similar concepts but from within the film-making industry. McIntyre describes Petrie's (1991) distinctions of creative agency "between the true 'auteur'– a film artist who works on material in such a way as to transform it into his or her own personal vision … and a craftsman [or craftsperson] who, no matter how skilled, can only adapt the concepts of others" as romantically inclined (As cited in McIntyre, 2012, p. 136). McIntyre goes on to establish that Csikszentmihalyi's (1997) systems approach is more suitable when discussing auterism because "apart from its refutation of the romantically inclined auteur theory [as above in Petrie (1991)] … it works for all the subfields within film-making" (McIntyre, 2012, p. 146). Importantly, the work of Anthony and Lefford (2019) also articulates a significant opposition to that of Petrie (1991). Anthony and Lefford (2019) suggest that all music producers can choose to move across the producer archetypes during the production of a song. Typically, such choices are determined by project requirements, the personnel, the musical content, possible genre constraints, etc. In these scenarios, producers sometimes use their agency, for example, to follow recognised conventions (extrinsic). Later they may act on their subjective preferences (intrinsic) to add distinctiveness. The work of Anthony and Lefford highlights that,

Table 2.1 Producer archetypes

Goal:	Intrinsic focus	Extrinsic focus	
Archetype:	Auteur	Innovator	Craftsperson
Value:	Subjectivity>novelty >recognisability	Novelty>recognisability =subjectivity	Recognisability>novelty >subjectivity

Source: Developed from Anthony and Lefford (2019).

no matter which archetype the producer operates within, they draw on all their past experiences including memories of music created by others, particularly those recordings that have been recognised and accepted in the domain as fine examples of a genre, style, technique, etc. The comments of producers in Anthony and Lefford (2019) make it clear that there is very little about professional production that is connected to a romanticised notion of creative practice, and this very much supports a domain-based PMP agency.

Music production cultures are influenced by the ways in which producers operate (archetypes) because, from within these approaches, producers adapt technological, musical and social activities. The producer archetypes can also act as *waymarkers*[2] for aspiring producers to use as summaries of practice or tools to unlock professional engagement. Phillip McIntyre discussed this approach with me, and he suggested that educators need to find ways to simplify some very heavy abstract concepts, particularly with undergraduate students. Phillip also outlined that danger manifests in students when one simplifies things too much and they start to believe the simplification is the real world. Regarding the producer archetype typologies, Philip suggested:

> If you say to your students … "here's the typology and this is what happens", that's when you run into trouble. But if you put the qualifier in front of it and say "here's a typology based on a particular set of cultural assumptions about the way creativity works, but it's useful to you, as a tool to unlock this world. Then … when you get into the world you're going to discover this typology has a lot of bleeding from one category to another. And, its not just that this person is locked into this one type, at various points they're going to be running across all three depending on what it is they are doing, who financed it, what it's for and what's going on at the time."
>
> (Phillip McIntyre interview, July, 2021)

The utility of the producer archetypes as a pedagogical tool is discussed regularly in the latter chapters of this book.

THE INFLUENCE OF CONSUMERISM

I had a fantastic conversation with Prince Charles Alexander during the creation of this book. He had some insightful information regarding the influence that commercialism and consumerism have on the production of music:

> For me billboard is like, ahh, the stock market, you know. So, there's the hot 100 charts…, how many country records have hit the hot 100 in the last 5 years? Very few. Okay, so if you wanna do country … let's go to the country chart. Boom. You get 'I Hope' by Gabby Barrett, right? 'I Hope' did hit the hot 100. Okay, what does it sound like? It sounds like a pop record. So, even your country is starting to have to

> come to terms with a lot of the concepts that we might discuss in the hot 100. You go to trap music. Okay, so trap music is doing a different thing. Everybody's rappin', everybody's kinda talk-singing. Alright, so there's a number one trap record. What does it sound like? It sounds like Dre. It sounds like a pop record! It's on the hot 100.... This is gonna be my philosophy.... I don't care if the rock n roll record sounds like an EDM record. I don't care if the R&B record sounds like a pop record. That's not the point to me. The point is how is music creating commerce. So, if you're a mixing engineer, you need to know how to mix this record. If you're a producer, you need to know how to compete with this record. Even if you don't wanna make that record, you still need to know how to be competitive with that record.
>
> (Prince Charles Alexander interview, April-May, 2021)

Consumerism can influence the way producers work because most productions are competing with others for popularity, sales, streams, etc. This brings with it possible influence from invested parties like recording and publishing companies. The producer may need to consider the wants of these invested parties and balance them with the artist's aesthetic intentions when producing. Industry influences on PMP, such as artist and repertoire people (A&R), were investigated by Negus (1992, 1999, 2002), and Frith, Straw, and Street (2001) discuss the impact that music consumption and markets, popular music and politics and race equality can have on popular PMP processes. What often becomes challenging in these landscapes for many aspiring popular musicians is that, in contrast to a commercial product, some see music making as a form of self-expression. This can be from an intrinsic perspective such as to protest, or to document events. Joe Bennett added some interesting context to this dichotomy when discussing undergraduate song subject matter:

> I think the biggest gap – between student creative activities in popular music degrees and the music that society consumes in the mainstream – is lyric theme? If you ask an 18 to 24-year-old what's important to them, with their particular social development in their lives at that time, they may not want to write something for the mainstream market, which is often slightly younger teens; they might not want to write about falling in love or unrequited love or heartbreak, which are the most common themes in most pop song lyrics. They think of themselves as deeper individuals than that emotionally, as we all do, because real life is usually more complex than can be covered in a three-minute pop song. So you end up with this mismatch, I think, between the emotional truth of a typical undergraduate – what they want to say about themselves and to the world – and the themes of the majority of all songs. I speculate that this is at least part of the explanation, at least when students are new to songwriting.
>
> (Joe Bennett interview, April-May, 2021)

There may be two or more reasons for producing music (self-expression, consumerism or a combination of these). These are dependent on the artist's aspirations – what does the artist wish to achieve with their music? With whom do they wish their music to connect and how should that connection be achieved? Once this has been established, the producer can tailor their skillset towards the particularities of the artist's vision and any relevant tastes and expectations of the intended audience.

CONSIDERING THE LISTENER, GENERATING AN EMOTIONAL RESPONSE

> Gradually it dawned on me how special it is to be able not only to capture sound – moments in time, ideas, concepts, personalities and attitudes – but to be able to fashion these elements into something that excites and inspires people all over the world.
>
> (Burgess, 1997, p. xii)

Burgess' words here are inspirational because they highlight that popular music producers are preparing an experience for others and their enjoyment. Producers consider who the end listener might be for every song with which they work, and they contemplate the ways in which they can improve the listening experience. When I interviewed Jan-Olof Gullö he was gracious enough to discuss with me his conversations with legendary pop producer Max Martin. One of the most instrumental things Jan-Olof and Max discussed was that producers need to understand:

> The fundamental ways in which people listen to music … if you are walking down the street and ask someone "what do you like about the song?" Many people will say "I like the lyrics". The lyrics are so important for the ordinary listener … much more important than the … drums or bass line … that we [producers] often focus on. [The public] won't hear if I change the bass drum a little bit.
>
> (Jan-Olof Gullö interview, April-May, 2021)

Listening to lyrics helps people to *connect* a part of their existence to the music: an emotion, a memory or something that they value. Different types of music use contrasting songwriting approaches, have varying sonic and instrumental characteristics and encompass a broad range of emotional experiences for those involved in the listening. These can be as diverse as the euphoric atmosphere associated with Rave Music (Siokou, 2002) and the socio-cultural connection to black culture that is associated with Rap or Hip Hop (Rose, 1994). Jan-Olof went on to attest to Max Martin's savant-like ability of knowing what people will love about a song and how to get the most out of that through the act of production. Jan-Olof closed by saying "when the producer, the musicians and artist can connect with the listener that is when magic is made" (Jan-Olof Gullö interview, April-May, 2021). The key takeaway here is how varied the listener's expectations might be.

Genre implies that the listener may have expectations about the product. Progressive rock and metal music have been through eras where the drum sounds were expected to be hard-quantised to a grid and the kick drum equalised to promote the hi-midrange frequencies. Bennett's research is illuminating in this area. His discussions on *What makes Trap ... Trap*? (Bennett, 2020) highlights the prolific use of the Roland TR-808 drum machine, the use of sub bass sounds, the importance of space and complex hi-hat and vocal poly rhythms in Trap music. Subsequently, Bennett (2018) suggests that there is a trend of popular music that is now tailoring to listener's preferences. For the producer, these considerations occur at the cognition stage, and continue through the production process and mixing. Nelson's (2016) "Between the Speakers" describes the author's approach to mixing as an intuitive process where ideas require plasticity so that the musical intentions of the song are nurtured. It is suggested that this process promotes emotional engagement on the part of the listener.

When considering how to develop these skills Prince Charles Alexander suggested aspiring producers can undertake sound alike applications where they attempt to imitate the popular works on the Billboard top 100. This process helps "connect to the pros and then they'll [producers] connect to the prospective audience" (Prince Charles Alexander interview, April-May, 2021). Gary Bromham suggested that producers need to consider how music is also consumed: "In education we really need to teach about the production-to-consumption chain.... Some of us are listening on YouTube ... my daughter, she's 12, she walks around and listens to music on her iPhone. She's not even listening in stereo!" (Gary Bromham interview, April-May, 2021). People listen to music in diverse locations, at home, in the car, on their laptop or phone and when listening many expect varying things from the listening experience: an emotional response or certain types of sounds, for example. Sören Johansson and I discussed that professional producers identify the *function* of the music: who is it intended for? How will it be consumed? What is it intended to do? After these qualities are identified, a hierarchy can be developed on production aspects that will aid in their expression so that the final result is tailored with the end listener in mind (Sören Johansson interview, April-May, 2021).

SOCIO-CULTURAL CONSIDERATIONS

The social environment of a music production culture is influenced by all members of the music-making team. The importance of the interaction between people involved in a session is integral. A crew mentality is often necessarily established: "making the best of your team to deliver on a project is part of the day-to-day work of the producer" (Hepworth-Sawyer and Golding, 2010, p. 77). Various people have the potential for both positive and negative influence on a music production culture due to their personalities, so producers often "meet the main artists in the project, [before recording] in person, in order to learn about their personalities, their artistic approaches and nuances" (Pras, 2016, p. 32) and this helps to manage the social dynamic of a session.

The impact of socio-cultural contexts on musical practice has been exemplified by the scholarly work of Cohen's (1991) study of the rock culture in Liverpool, England. Cohen (1991) explains:

> Individuals produce and consume music within specific social contexts (households, neighbourhoods, etc.); at specific times or historical moments; within specific networks of social relationships (involving kin, peers, colleagues, etc.), relationships that have different dimensions (social, political, economic). People's experiences of music, the uses they have for it, and the meaning they construct around, or through it, are bound up with these specificities, and with the interconnections between them.
>
> (p. 135)

Similarly, all of the characteristics described in Cohen's quote can be contextualised within a music production culture. Three key considerations are: the relationships of individuals involved, the historical times the music was produced and how this influenced discussions and engagement, and, most importantly, the individuals' experiences during the music making. However, music production currently resides in producer-led, co-produced, producer and band and self-produced landscapes, and these different *modus operandi* can influence the social nature of a session. Moreover, popular musicians also collaborate online and via virtual means. In these situations, communication often occurs within virtual chat rooms, networking can occur on social media platforms and the applicability of communication has been revolutionised by the virtual world. The key take away here is that PMP includes a diverse range of social realities that are influenced by the people involved, their unique personalities and the evolving nature of PMP as a form of practice.

Creative conflicts

The seasoned producer draws on psychological skills to nurture the people involved in the studio and this nurturing can take many forms. Being friendly, joking around, showing empathy or sympathy, being assertive, or even being confrontational can help to direct the music production culture in the right direction. Creativity and conflict go hand in hand. As Morrow (2012) suggests "it is clear that a certain degree of conflict within creative groups is necessary for artistically successful outcomes" (para. 36). When these creative conflicts occur, the experienced producer realises that the best way forward, in group situations, is to continually make decisions, even if those decisions end up being wrong. As Hennion (1989) states, "studio work proposes equivalences that are guesstimates, trials that work in the midst of many failures" (p. 420). The creative conflicts that occur during music making are necessary processes the production team will go through to filter out any problems in the music, but these are not always easily traversed. Accordingly, producers need to "prepare to experience problems with multiple egos and the subjective opinions of the performers" and focus on the development of "respectful personal relationships [which] contribute to creating a successful project" (Zager, 2006 p. 123).

My traineeship days provided many instances of creative conflicts and these experiences re-affirmed to me the importance of communication and psychology in PMP. One occasion involved an extremely high-profile American producer (identified here as Pete) during the recording of what would become a multi-platinum album for an iconic Australian artist. Pete was working with the band, trying to record and finalise the drum tracks for the album so overdubbing could occur. The drummer for the session was a renowned musician in Australia (identified here as Scotty) and I remember that, on that day, Scotty was struggling to satisfy Pete with his performance for one of the songs. Pete was giving him advice and being proactive, but he eventually suggested he wanted to play the drums and give Scotty an idea of what he wanted. I don't think anyone (except for Pete) knew he played the drums. Regardless, he went into the recording room, took the drums sticks from Scotty, the tape rolled and Pete subsequently had a run through playing the drums to the track. When Pete walked back into the control room, we all realised Scotty had left the building. Ensuing negotiations took place to get Scotty back on the session, but he did not return and the remainder of the drums were tracked in Los Angeles.

This story exemplifies that music making encompasses difficult aesthetic and musical decisions which must be confronted. If a drummer's performance is not up to scratch, producers need to communicate with them so as to help them improve or find another way. From the outset, I often say to my production clients:

"You can rest assured that you will get honesty and empathy with me, but sometimes it is going to hurt. Just don't forget: it's not personal; its business. You have hired me to do a job. My job is to make it right. It's all for the betterment of the music."

However, the ways in which producers communicate these things are of the highest importance because the social landscape within which musicians work is extremely complex and involves the interaction of diverse personalities.

Another example I remember was when a legendary producer from the UK (identified here as Mark) was working with an esteemed harmonica player from the States (identified here as Willie) during the production of a world-renowned Australian band's multi – platinum selling album. Whilst Willie was noodling through some parts, Kurt (songwriter, keys, guitar for the band) also started suggesting some ideas. This did not go down well with Mark who glared at Kurt in insistence he leave the producing to him. As far as I know, the relationship between Mark and Kurt was very strong, but, at times, strong personalities clash. I remember a conversation that occurred one morning after the "Willie" conflict. Kurt entered the control room and asked Mark if he would be needed for anything today. Mark informed him that he wasn't sure. Kurt then replied: "Well, contrary to what you might think Mark, my time is quite important to me". Mark retorted with "well fuck off then!" Kurt returned fire: "no, how about you fuck off!" And Mark did. He walked out of the session and, I believe, headed up Oxford Street to have a coffee. Ensuing panic took over; band members talked to Kurt suggesting that it was probably not a good idea to tell the producer to fuck off and that he needed to apologise. To his credit, Kurt did apologise and the two came together, made amends, and the session moved forward. The tension between Kurt and Mark manifested from a leadership dispute, namely the producing of Wille. However, it just goes to show that the pressures associated with making music – the expectations of the record company and the public, and the overarching impact music making can have on our personal lives (time with family, etc.) – have the potential to infiltrate and exacerbate any tensions during a session. This further emphasises the importance of a producer's communication skills, and this is transferable to all forms of music production: acoustic, electronic, etc. As my colleague Annelise Noronha suggested to me: "the one thing that stays the same [across all forms of PMP] is a producer's *bedside manner*. That stays the same and is really important" (Annelise Noronha interview, April-May, 2021).

To the novice, it may feel more comfortable to avoid creative conflicts, but popular music is not made this way. One cannot ignore a musical *faux pas* for the sake of relationships. Producers must aspire to generate great performances and hunt down and eliminate any detracting musical or sonic elements. However, they must also realise that they are working with people, social beings who have unique personalities, attributes and feelings. Therefore, the relationships between those making the music must be nurtured whenever possible.

Forming of relationships, networking, ego, psychology and communication

In June 1990 whilst in London mixing INXS's album *X* (Farriss et al., 1990), Chris Thomas, Dave Nicholas (engineer) and I went to an outdoor concert at a place called Knebworth. Chris took me backstage and introduced me to the who's who of English rock n roll, many of whom he had produced. Later that night, whilst drinking, I recall a conversation we had where Chris highlighted to me that it was the temporary relationships formed with artists that make a bond; the glue that holds the project together. By going backstage at Knebworth and saying hi, we were nurturing those relationships. These networking moments are valuable because when people are *in the trenches* making music together they help provide the trust required to traverse musical conflict for the betterment of the music.

One of the overarching issues with PMP social situations is that popular musicians need to be confident decision makers and have belief in their ability and inclinations. However, there is nothing more damaging than an out-of-control ego on a session. During PMP, individuals need to be okay with having their performances critiqued since it is an immutable part of the process. The age-old adage, *leave your ego at the door*, presents a necessary skill for all involved in the process. In recording situations, some aspiring professionals feel they need to prove their worth by identifying every incorrect element in a musician's performance. However, sometimes it is more useful to be strategic and help the musician work this out themselves. There is often little gained by an aspiring engineer exemplifying the expertise of their musical ear if it detracts from a singer's confidence. Conversely, if a singer recognises the area of their performance that is incorrect and identifies that to all in the room, this empowers the vocalist and they take ownership of their performance and become accountable, and this nurtures the singer's engagement. Klein (2013) recalls the traits of a New York-based guitarist who was always busy and employed regularly:

> He shows up early, gives it his all, and always plays great stuff. Oh, and there's one more thing. The guy is a blast to hang out with. He's easy going and funny and generally a good hang without being distracting... The work gets done without any problems, sounds great and everyone goes home happy.
>
> (p. 33)

Within music making scenarios, it is important that popular musicians understand who they are as social beings and what they bring with them to a production setting. Producers like Rick Rubin have been known to proclaim the importance of developing a spiritual focus on one's own sense of self (Hirschberg, 2007). These are some of the necessary psychological concepts popular music producers use to influence the socio-cultural climate within which they work.

Music production lexicon

Moylan (2002) states that "a vocabulary for communicating specific, pertinent information about sound quality does not exist. People have been talking about sound for hundreds of years without a vocabulary to describe their actual perceptions and experience" (p. 87). Although I generally agree with Moylan's comments, it is important to understand that popular musicians converse with and understand each other in specific ways. PMP engages with more than just sound; it both informs and is informed by the musical and social culture that surrounds it (Zak, 2001). As such, the lexicon and topics of conversation that musicians use during recording influence the music production culture. Howlett (2007) discusses a paper presented by Thomas Porcello at the Art of Record Production (ARP) conference, Edinburgh University (2006), in which Porcello "reported on his year hanging out in studios recording the terminology and referential discourses used by producers, engineers and artists to articulate the kinds of sounds and musical parts desired" (as cited in Howlett, 2007, p. 2). Porcello's (2004) publication delves deep into the way recording engineers attempt to express themselves so that musicians, producers and composers understand the deeper musical or sonic connotations of the topic. His study illuminated the:

> ...enormous subtlety of the multiple ways in which music is talked about in studios, and the importance of both acquiring a knowledge of this range and of being socialized into the proper use of a sound

> engineer's discursive practices.... Without knowing the rules that govern the proper use of the linguistic resources ... one will remain marked as an outsider.
>
> (pp. 753–754)

These rules that Porcello describes are all informed by the music production culture. They are the specifics of the music, the technology, the creative practice and the skills being used in the session. Being an informed popular musician in these areas seems integral for useful communication and promotes PMP being carried out from a both-sides-of-the-glass rationale. In this environment, musicians need to be able to converse regarding technological applications and engineers and producers need to be able to converse on musical and creative levels.

By engaging in diverse communicative processes, popular musicians become skilled in their application and learn to intuitively adapt these depending on the people with whom they are working. A useful tool in this context is humour. Humour is often used by producers to relax performers, and this nurtures the atmosphere during recording and promotes higher quality performances. Mixerman (2012) writes "as much as making a record is serious business, the goal is to make it a fun experience" (p. 286). During a production, musicians and producers will also have favoured topics of discussion: the food they are eating, the sport they like watching, the current state of their love life and perhaps the latest band they went to see. All of these conversations inform the music production culture and, at times, these conversations engage with profanity.

When reading professional texts like Mixerman (2010), expletives appear frequently. This is not because the author is trying to be offensive, it is because of the culture within which he has been immersed as a popular music producer – curse words are a part of his lexicon. However, the measure of these words is often a result of the people engaged in the session and the producer will adjust to the people with whom they are working. Music production cultures (like many cultures) are reactive to the people involved in them. For example, I have worked on many religious recordings where swearing was definitely not appropriate, and consequently I adjusted my lexicon so as not to use expletives. In comparison, some musicians use swear words as a form of bonding. In my experience, this has included Punk, Folk, Progressive Rock and Hip Hop artists amongst others. Therefore, even though swearing is not ubiquitous across PMP practices, sometimes a producer can develop relationships by using a lexicon that the artist engages with and understands.

Throughout my years as a popular music producer working with internationally acclaimed producers and artists, swearing was a constant part of the culture. Swearing seemed to lighten the mood at times; it diminished the stress regarding the successful expectation of the music and it relaxed the atmosphere in the studio. It helped all involved to feel like we were on the same team. There is no doubting that this is a problematic situation in an educational context, especially given the current climate and strict policies around workplace language. Yet it is clear that the use of expletives is common in the professional industry, and negotiating with this is accordingly necessary in pedagogical contexts. The lexicon used within PMP creative practice helps the participants achieve the ultimate goal – a high-quality musical output. How it does this is by nurturing what popular musicians commonly refer to as *the vibe*.

THE VIBE

Zak (2001) contextualises the importance of the "atmosphere in which recordists work and which influences their creative imaginations and aesthetic decisions" (p. 184). Zak tables terms such as *convention*, *association*, *allusion* and *rhetoric* to discuss "the workings of an aesthetic system whose very nature shapes the production and reception of its constituent works" (p. 184). Zak's use of the phrase *aesthetic system* connotes a kind of creative space that links to discussions by Minchella (2018), who uses the word *atmosphere*. These ideas can all be related to an important and commonly used term in popular music professional practice: *the vibe*. Siokou (2002) uses the word vibe in her discussions of the Melbourne rave scene. In this context the word is used to describe a collective "party atmosphere" (p. 13), and the author suggests the vibe helps to unify the crowd so they feel "emotionally part of something larger than themselves" (p. 11). Hepworth-Sawyer and Golding (2010) proclaim:

> ...many engineers and producers we have spoken to have mentioned the word vibe. The problem with getting the vibe right is that it really varies depending on so many different factors.... It's whatever works in order to get the best take possible; the key is to be perceptive, open minded and creative.
>
> (p. 225)

The vibe is an integral element of the PMP process and my industry experience informs that it is a result of many factors. These factors can be strategically planned and can include having snacks and drinks available or engaging with social elements such as relaxed discussions at pubs, cafes and restaurants. There are also numerous recreational activities like gaming[3], engaging with social media or the Internet, playing table tennis, table soccer, cards or going out to clubs or gigs.

The recording may also take place in a residence. These lived experiences stimulate emotions and feelings in musicians whilst working (Minchella, 2018). They inspire relationship-building conversations that can be continued during recording and this develops trust between musicians. The vibe of a performance space, as discussed by Gibson (2005), can also be manipulated by the dimming of lights (Mixerman, 2012), the use of humidifiers, or the hanging of ornaments, and this helps to ensure that a comfortable space is available in which to perform. All these elements help to distract the musician from making poor musical decisions; they temporarily clear the mind of unnecessary and unhelpful thoughts and promote clarity and the potential way forward for the recording.

THE IMPACT OF THE MUSIC ON MUSIC PRODUCTION CULTURES

Seabrook's *The Song Machine* (2015) explores the varying forms, and respective processes, of popular music hits. It is an insightful read into the commonalities and idiosyncrasies of different types of popular music. Seabrook describes US rapper Eminem's music as "angry, profane and misogynistic [a mean voice for a dark time]" (p. 121) but also describes the highly predictable and stringent musical rules related to many "Hit Factories" such as K-Pop and the "girlish quality" (p. 87) of Britney Spears' music. These different types of music require the use of specific musical techniques (rapping, manipulating beats, four-part vocal harmonies), they require suitable forms of technology (tape machines, sequencers, computers, electric guitars) and are created in varying social atmospheres with people who use diverse lexicons and have unique social behaviours. Therefore, the characteristics of the music being made influence the makeup of the music production culture itself.

PMP is also sometimes influenced through an engagement with society and the cultural realities of our world. Regarding Rap music, Rose (1994) writes:

> Rap's black sonic forces are very much an outgrowth of black cultural traditions.... Many of its [Rap's] musical practitioners were trained to repair and maintain new technologies for the privileged but have instead used these technologies as primary tools for alternative cultural expression.
>
> (p. 63)

Music making is at times a reflection of, or reaction to, how musicians live their lives. Their expressivity is important and this comes out in the idiosyncratic and recognisable way they dress and communicate, and even in their body language. These influences are fundamental to the ways music is made and as such need to be deeply considered when planning and undertaking PMP.

SUMMARY

I have outlined in this chapter many of the cognitive and socio-cultural realities that occur in modern PMP. The importance within the processes listed above is that they inform and influence the manifestation stage of music making (discussed in Chapter 3). Therefore, an engagement with them is integral for an aspiring producer's education. Later chapters of this book will discuss how pedagogical frameworks can facilitate an authentic learning experience in these concepts through an experiential process that integrates critical thinking and practical engagement within a HE context.

REFERENCES

Anthony, B. (2017). *The producer's vision: A study into the multi-faceted cognitive design of the popular music recording aesthetic.* Paper presented at the 12th Art of Record Production Conference, Royal College of Music Stockholm.

Anthony, B., & Lefford, M. N. (2019). *The producer's vision: A study into the multi-faceted conceptions of the popular music recording aesthetic.* Unpublished article.

Bennett, J. (2018). Songwriting, digital audio workstations, and the Internet. In N. Donan (Ed.), *The Oxford handbook of creative process in music* (pp. 1–30). New York: Oxford.

Bennett, J. (2020). What makes Trap… Trap? Retrieved from https://tidal.com/magazine/article/what-makes-trap-trap/1-72722

Burgess, R. J. (1997). *The art of music production* (3rd ed.). London: Omnibus Press.

Burgess, R. J. (2013). *The art of music production: The theory and practice.* Oxford: Oxford University Press.

Cohen, S. (1991). *Rock Culture in Liverpool: Popular music in the making.* Oxford: Oxford University Press.

Covach, J. (2006). From "Craft" to "Art": Formal structure in the music of the beatles. In K. Womak & T. Davis (Eds.), *Reading the beatles: Cultural studies, literary criticism, and the fab four* (pp. 37–53). New York: State University of New York Press.

Csikszentmihalyi, M. (1997). *Creativity: The psychology of discovery and invention.* New York: Harper Perrennial.

Draper, P. (2013). On critical listening, musicianship and the art of record production. *Journal on the Art of Record Production,* (8). Retrieved from http://arpjournal.com/2630/on-critical-listening-musicianship-and-the-art-of-record-production/

Farriss, A., Farriss, J., Hutchence, M., & Thomas, C. (1990). *X* [CD]. [Recorded by INXS]. Sydney: Atlantic records.

Foo Fighters. (2011). *Wasting Light* [CD]. California, U.S.A. RCA.

Frith, S., Straw, W., & Street, J. (2001). *The Cambridge companion to pop and rock.* Cambridge, MA: Cambridge University Press.

Gibson, C. (2005). Recording studios: Relational spaces of creativity in the city. *Built Environment, 31*(3), 192–207. DOI: 10.2148/benv.2005.31.3.192

Hennion, A. (1989). An intermediary between production and consumption: The producer of popular music. *Science, Technology, & Human Values, 14*(4), 400–424. DOI: 10.1177/016224398901400405

Hepworth-Sawyer, R., & Golding, C. (2010). *What is music production? A producer's guide: The role, the people, the process.* Burlington, MA: Taylor & Francis.

Hirschberg, L. (2007, September 2nd). The music man. Interview with Rick Rubin. *The New York Times* (online), Retrieved from https://www.nytimes.com/2007/09/02/magazine/02rubin.t.html

Howlett, M. (2007). *Fixing the volatile studio: Studio vocal performance techniques.* Paper presented at the 3rd Art of Record Production Conference, Queensland University of Technology, Brisbane, Australia. Retrieved from http://eprints.qut.edu.au/33275/

Klein, Jim. (2013). *Welcome to the jungle: A success manual for music and audio freelancers.* Milwaulkee, WI: Hal Leonard Press.

Marrington, M. (2016). Mixing metaphors: Aesthetic, mediation and the rhetoric of sound mixing. In R. Hepworth-Sawyer & J. Hodgson (Eds.), P*erspectives on music production: Mixing music* (pp. 199–215). New York: Routledge.

McIntyre, P. (2012). *Creativity and cultural production: Issues for media practice.* Basingstoke: Palgrave MacMillan.

Minchella, D. (2018). The poetics of space: The role and co-performance of the spatial environment in popular music production. In S. Bennett & E. Bates (Eds.), *Critical approaches to the production of music and sound* (pp. 41–61). New York: Bloomsbury.

Mixerman. (2010). *Zen and the art of mixing.* Milwaukee, WI: Hal Leonard Books.

Mixerman. (2012). *Zen and the art of producing.* Milwaukee, WI: Hal Leonard Books.

Moll, James. Director. (2011). *Foo fighters: Back and forth.* Film (documentary). California, U.S.A. Exclusive Media Group, RCA Records, Back & Forth Pictures

Moorefield, V. (2005). *The producer as composer: Shaping the sounds of popular music.* Cambridge, MA: MIT Press.

Morrow, G. (2012). Creative conflict in a Nashville studio: A case of boy and bear. *Journal on the Art of Record Production, 6.* Retrieved from: https://www.arpjournal.com/asarpwp/creative-conflict-in-a-nashville-studio-a-case-of-boy-bear/

Moylan, W. (2002). *The art of recording: Understanding and crafting the mix.* Boston, MA: Focal Press.

Negus, K. (1992). *Producing pop: Culture and conflict in the popular music industry.* London: Edward Arnold Publishing.

Negus, K. (1999). *Music genres and corporate cultures.* London: Routledge.

Negus, K. (2002). The work of cultural intermediaries and the enduring distance between production and consumption. *Cultural Studies, 16*(4), 501–515, Retrieved from https://doi.org/10.1080/09502380210139089

Nelson, D. (2016). Between the speakers: Discussions on mixing. In R. Hepworth-Sawyer & J. Hodgson (Eds.), *Perspectives on music production: Mixing music* (pp. 122–139). New York: Routledge.

Petrie, D. (1991). *Creativity and Constraint in the British Film Industry.* London: Palgrave MacMillan.

Porcello, T. (2004). Speaking of sound: Language and the professionalization of sound-recording engineers. *Social Studies of Science, 34*(5), 733–758. DOI: 10.1177/0306312704047328

Pras, A. (2016). What has been left unsaid about studio practices: How producers and engineers prepare, manage and direct recording sessions. In A. King & E. Himonides (Eds.), *Music, technology, and education: Critical perspectives* (pp. 27–44). New York: Taylor and Francis.

Prime loops. (2012). Inside the World of 90s Pop. [Onlilne]. Retrieved from https://primeloops.com/blog/blog/post/slug/90s-pop-what-can-britney-teach-us-about-music-production/

Rambarran, S. (2015). *"'99 Problems' but Danger Mouse IS One": The evolution of the Producer Brian Burton.* Conference paper presented at the 10th *Art of Record Production Conference*, Drexel University, PA. Retrieved from author.

Ronson, M., & Remi, S. (2006). *Back to Black* [Recorded by Amy Winehouse] [CD], U.S.A.: Island.

Rose, T. (1994). *Black noise: Rap music and black culture in contemporary America.* Hanover, NH: University Press of New England.

Schippers, H. (2007). The marriage of art and academia: Challenges and opportunities for music research in practice-based environments. *Dutch Journal for Music Theory, 12*(1), 34–40. Retrieved from https://core.ac.uk/download/pdf/143877784.pdf

Seabrook, J. (2015). *The song machine: Inside the hit factory.* New York: Norton.

Shepherd, I. (2011). *Foo Fighters-Recording and Mixing 'Wasting Light'*. Retrieved: https://productionadvice.co.uk/foo-fighters-recording-wasting-light/

Siokou, C. (2002). Seeking the vibe [An analysis of the Melbourne rave scene.]. *Youth Studies Australia, 21*(1), 11–18.

Suddath, C. (2011). *Music Monday: The musical influences of Amy Winehouse.* Retrieved from http://newsfeed.time.com/2011/07/25/music-monday-the-musical-influences-of-amy-winehouse/

Thompson, P. (2019). *Creativity in the recording studio: Alternative takes*. Cham: Springer.

Zager, M. (2006). *Music production: A manual for producers, composers, arrangers, and students*. Lanham, MD: Scarecrow Press.

Zak, A. (2001). *The poetics of rock: Cutting tracks, making records.* Los Angeles, CA: University of California Press.

NOTES

1 Hybrid approaches integrate both acoustic and electronic music production techniques.

2 A *waymark* is a series of signals or signpost that mark out the route of a journey.

3 Gaming is the activity of playing video games on computers or consoles such as the Xbox One X and PlayStation 4.

3

Popular music production processes (Part B). The manifestation of recorded works

> Music Production is no-longer a siloed skill to be taught in isolation of the practices of songwriting and performance. By that I mean, production and the manipulation of digital audio is now mostly embedded with the songwriting and performance process and is less and less a separate practice.
>
> (Lachlan Goold Questionnaire reply, May, 2021)

MODERN PRODUCING, RECORDING AND MIXING

Lachlan's comments above reaffirm that during practical activities, modern PMP can utilise traditional frameworks (collaborative engagements between many individuals who are specialists in the various skills of PMP) or, in contrast, include an integrated approach where multidisciplinary practitioners bleed across the roles of songwriting, performance and production (this latter alternative is perhaps the more common of the two). Much of a producer's agency is influenced by the song. Therefore, in the following chapter (Chapter 4), I will unpack the relationship between songwriting and production and the diversity associated with songwriting as a form of creative practice. In this chapter I will outline many of the concepts, applications and influences common to the practical stages of PMP.

Pre-Production

Pre-Production is a foundational part of the production process that can occur before recording, or during songwriting and production. Its situation is dependent on the producer's approach. Pre-production has traditionally been used as a "planning phase before recording begins. Includes budgeting, production planning, scheduling, studio bookings and equipment rental, but it mostly refers to the selection, rehearsal and refinement of the music and musical arrangement prior to recording" (Burgess, 1997, pp. 297–298). This method of pre-production is utilised for acoustic or hybrid recording processes, and in this scenario the band/artist will meet with the producer and rehearse the songs to be recorded. During this time, all people involved will communicate about the song's melody, lyrics, key, tempo, structure,

DOI: 10.4324/9781003055099-5

arrangement and instrumentation. Skills in envisioning are especially important in this phase since choices made can have long-term effects on the production's course. Without vision, pre-production can become more like a jam session. This is not an entirely useless process if new grooves, ideas or songs need to be developed. However, pre-production is better suited if a context is understood and aspired towards by all involved. The pre-production process can be recorded to trial suitable technological setups and playback allows critical reflection and further adaption to instruments sounds, parts and lyrics.

Some artists who intend to utilise acoustic/hybrid approaches may prefer not to run a rehearsal type pre-production scenario. Instead, they may opt to employ the flexibility of the DAW to demo a version of the song at their home studio. Virtual drum kits, amp simulators and instruments can be used to do this. This method allows the producer to hear all of the parts working in a musical context, to listen to the song's structure, lyrics, tempo, key, etc., prior to applying explicit attention to committed sounds and performances for the final recorded product.

Pre-production can also occur *during* the songwriting and production process. Bennett's (2018) study of songwriting found that "the DAW's layout encourages certain creative behaviours that draw the songwriter away from traditional linear lyric/melodic composition and towards a more production-centred 'vertical' approach" (p. 8). An example of this is the tracker/top-liner process which is an innovative way of making music within the DAW. I have described this process in Chapter 1, but fundamentally the tracker programs a backing track to which the top-liner writes a melody and lyrics. Sometimes this process happens in reverse order or via a back-and-forth process, but the basic principle remains. In this context, pre-production occurs at the same time as songwriting, programming and production. The tracker/top-liner process blends these elements into a fluid creative practice where each element is informing all others, and all are reactive to the aesthetic decisions made by the producer.

The front end of music production (musical instrument knowledge)

Most students spend a lot of time trying to fix things in the mix. To avoid this, professional practitioners do a massive amount of work on what is commonly known as the *front end* of a recording. This requires the integration of envisioning and the explicit tailoring of a sound before it is committed to the recording. Therefore, music producers benefit greatly by having extensive knowledge of musical instruments of all kinds and how they produce sound. Some examples of this knowledge, pertaining specifically to the drumkit, could include an understanding of build and construction choices (drum size and materials used), the utility of dampening and tuning of the drum kit, different cymbal types and the use of sticks, rods, brushes or mallets (see Toulson, 2021). Moreover, this type of knowledge, as applied to electric guitar, could include knowledge of different types of pickups (materials, active vs passive) and how the position of these affects the sound. Producers benefit from a fundamental understanding of the instrument – it is helpful to know the difference in tone between a Fender Stratocaster (with single coil pickups) and a Gibson Les Paul (with humbuckers). Producers should understand the difference between hollow and solid body instruments, the specifics of tuning on a five-string bass guitar as compared to a four-string instrument, and so forth. They should know the various tunings that guitarists can use (open, standard, etc.) and the different sound sculpting capabilities of the plethora of guitar effect pedals and amplifiers that are available. Moreover, knowledge in vintage keyboards like Hammond organs, Leslie speaker cabinets, Wurlitzer Pianos, Fender Rhodes and synthesis, sampling, found sounds, warping (time stretching) autotuning and electronic music processes, etc., is always useful. Additionally, any knowledge in classical and world music instruments is advantageous. Finally, and most importantly, producers benefit if they have some knowledge of vocal anatomy and how vocalists can produce varying sounds like head voice/falsetto, vocal scraping, vocal fry, chest voice or belting, for example. This knowledge does two things: firstly, it facilitates efficient and effective communication between the producer and the various performers they may work with, and secondly, it commits sounds throughout the recording process, which integrates the producer's instincts, intuitive decision-making process and engagement with vision. Attention to the front end of a recording also minimises the need to fix things when it comes to mixing.

Musicality and performance

Musicality and performance are often the PMP skills most undervalued by students. Music is made through the performative abilities of the people that are involved with the project and high-quality musicians can make an enormous difference to the musical work. Musicality and performance approaches can be diverse, but, within PMP, this would include traditional musical understanding of chords, modes, arpeggios, scales, melody and rhythm. It would also integrate performative concepts of feel, dynamics and a sense for suitable tempo or meter that may be stringent or fluid. High-quality musicians consider the context in which they are playing; they play only what is needed (they don't over play), they improvise when needed or use repetition to accentuate hooks. Performance is often driven by what musicians hear as suitable or by their intention within the song's context; they may aspire to blend with others within a group setting but also seek contrast in their playing when their part needs to shine. Good performers always value the spaces that are left (when playing does not occur) and are meticulous regarding the maintenance of their equipment and the suitability of the instrument they are using so that performative goals can be met. These are elements producers seek to target when generating performances and are useful considerations within production discussions.

PMP also brings with it new types of performative applications. For example, playing synthesisers and adjusting the envelope, LFO and filters in real time. Many DAW-based musicians integrate the laptop with some form of MIDI controller during performances, for example, a MIDI keyboard, drum pad or third-party performative device like a *Novation Launchpad, Ableton Push* or *Native Instruments Maschine*. These devices all require an explicit understanding of the instrument and suitable skill to perform on the device within the production process.

The traditional recording studio has, itself, been employed as an instrument. This idea originates from works such as Kealy (1979, 1982) where the recording studio is described as an extension of the music producer's creative arsenal: "The rock musician views the studio equipment as practically another instrument. As a result, the occupational ideology for sound mixing changed: work previously considered merely technical now became artistic" (Kealy, 1979, p. 18). Schmidt Horning (2012) also suggests that "in the process, listeners' expectations have been shaped by many factors, but none more than the changing uses of the studio as instrument" (p. 41). Mixerman (2010) writes "mixing is neither a linear process nor a technical one. It's a musical process, and as such, a mix is something that one performs – like an artist" (p. 15). I wrote an article in support of this theory where I discussed a concept that originated from the pre-automation days or an *All-hands-on-deck* approach (see Williams, 2011). In these scenarios "mixing as a performance involves a creative practice where the recording studio is the mixer's instrument; during this process, audio technology [faders and equalizers are] manipulated by the mixer in a performance-like process to create desired artistic outcomes" (Anthony, 2017b, p. 1).

The influence of the "spaces" and "places" of PMP

I have recently written a chapter on the spaces and places of PMP (Anthony, 2022), and I acknowledge here that much of the content within this section has been taken from this resource and paraphrased. Tuan (1979) posits that place and space often merge and are interrelated where "'space' is more abstract than 'place'" (p. 6) because place reflects "centres of felt value" where human needs are satisfied (p. 4). Tuan suggests the correlation between the two is that "what begins as undifferentiated space becomes place as we get to know it better and endow it with value" (p. 6). Within the agency of PMP, place and space are interconnected and, moreover, they have fluid meanings that link cognitive processes with practical applications.

Place, within the realms of PMP has many meanings. A known or familiar place can become realised within a certain recording studio and its technological affordances (Watson, 2015) or by applying "specific equalization to the headphone mix, and adding reverb and delays to create an aural atmosphere the singer feels comfortable with" (Howlett, 2009, p. 32). Place can include the considerations of a geographical position (Stockholm, for example), these places have distinguishing factors like rolling hills, buildings, nightlife, food, sport and culture. These things inspire conversations and may influence the music that is written or performed. Since the invention of the DAW, small-scale and home studios are being used ubiquitously. These places have unique equipment affordances (audio technology) and room dimensions, and the architecture is considerably different to large studios. Picture yourself working in Air Studios London in comparison to a 21-year-old Hip Hop producer's lounge room in New York City. Both these places influence the music-making process and, as such, producers consider the affordances of the places within which they work most ardently. Metaphorical place is also interesting to consider (see Walther-Hansen, 2020). Lyrics that describe a "cold dark room", for example, can represent an introspective point of view. Sometimes it is important to ask performers to take themselves to these metaphorical places when they perform so the correct emotion is generated in the performance.

Space, as mentioned above, is more abstract. *Technospheres* are "an imagined space of communicative potentialities and constraints" that, through technology, locate and ground a producer (Toynbee, 2000, p. xxii). These potentialities and constraints are fundamental to whether a producer will select a given piece of technology. Consider the different technospheres between an analogue tape machine and a cloud collaboration using Pro Tools. There are significant musical, technological, communicative and social potentialities and constraints that are associated with both technologies. This will influence the musical work and, as such, must be considered by the producer during envisioning and planning sessions.

Some of the most challenging work producers do is with a person's *headspace*: their state of mind, thoughts, feelings and emotions. Howlett (2009) suggests "the arts are renowned for tales of irrational, obsessive, even abusive and neurotic behaviour. My observation is that it is almost integral to the function" (p. 29). Performers need to be in the right headspace to perform and the producer must consider many variables that may influence this. Has the performer eaten and had enough sleep? Is it the right time of day to perform? Are they capable of taking themselves to the necessary emotional headspace during performance? These matters are informed by the song's message and can require a broad scope of emotions from frailty, anger, love or happiness. There could be sexual connotations or the use of profanity in the lyrics of the song that require consideration. This is where place and space interconnect and the producer's job is to position the performer in a suitable place that enables them to interact with the necessary headspace and generate an appropriate performance for the song. Obviously, there is a lot to consider here with particular regard to pedagogical landscapes, but professional practice dictates that matters of place and space are an integral element of PMP.

Sonic textures (advanced concepts of sound recording) and production techniques

My university and industry training allowed me to discover the utility of sonic textures and advanced sound engineering/production techniques and that these only complimented my prior songwriting and performance skillset. Like many of my students, there was a time when I just loved fiddling with gear and experimenting with different microphones and microphone techniques. I enjoyed audio sculpting through the use of panning, effects, distortion, distinctive equalisation, compression and layering sounds whilst in search of that sonic "wow factor". I fear that in my emphasis on the socio-musical applications of PMP I have de-valued the importance of such practices. To be clear, the considered use of sonic textures (and the techniques that generate them) to support musical intentions and elicit an emotional response from the listener is not only fun, but incredibly important. I am not going to go through the many processes one can apply here; that is tacit knowledge that is accomplished by research, practice, intuition, experience and a love for listening to PMP in general. I will say, however, that there are great resources online that unpack some of the incredible work of extremely accomplished professionals. Beato (2018) [see link in reference section] is a shining example of the exemplary programming, musicality, performance, sonic layering and production processes utilised by Max Martin on Ariana Grande's hit single "Into You" (Grande et al., 2016). This track exemplifies some of the applications I have listed above but also illuminates the incredible things a producer can achieve with considered use of double tracking, harmonisation, call and response, multiple musical hooks and incredible vocal performance and production techniques. This track makes you want to move, gives you chills and inspires you to keep making music.

Mixing and mastering

Mixing has been researched, defined and had its myriad of secret practices illuminated for decades. At times, if you look on the Internet at various audio chat rooms like Gearspace (2020), it is as if great mixers are akin to musical Jedi Knights. They are revered and seen as incomparable in their approach versus the aspiring producer. However, a suitable definition of what mixing is suggests:

> A process in which multitrack material – whether recorded, sampled or synthesisized – is balanced, treated and combined into a multichannel format. Most commonly, two channel stereo. But a less technical definition would be: a sonic presentation of emotions, creative ideas and performance.
>
> (Izhaki, 2008, p. 5)

Izhaki's quote represents a common disparity in both research and practice when discussing mixing. There appears to be two distinctive considerations: technical (sonic) and musical (creative). In my opinion, and it is probably of no surprise given what I have written earlier, the technical side can be learnt and exercised through continued practice. The greater skill in mixing involves an acute awareness of critical listening which allows the producer to work out the function of the song and then promote any musical, performative and sonic elements that support this. Izhaki writes, "It is for their sheer creativity – not for their technical brilliance – that some mixes are highly acclaimed and their creators deemed sonic visionaries" (2008, p. xiv).

There is a lot of scholarly work on mixing. A good resource that employs a number of professional practitioners as authors is *Perspectives on Music Production: Mixing Music* (Hepworth-Sawyer and Hodgson, 2016). Within this text, Harding (2017) tables a pragmatic approach to "Top Down Mixing", a common technique of the 1980s where the mixer attended to the lead vocals first before making their way through the arrangement to the drums. Paterson (2016) illuminates the diversity in mixing practice catalysed by the advent of the DAW. He suggests that there are now *overlaps* where the mixer is, in fact, capable of production-based decisions and processes. Relatedly, Lauzon (2016) unpacks the relationship between pre-production and mixing. In this work Lauzon describes his practice of recording the entire song during pre-production which forms a mix-forward perspective to PMP.

Moylan's (2017) chapter discusses the tacit listening practice of professional mixers and suggests that mixing is a personalised process where professionals learn *what* to listen for (individual sounds or a whole mix) and listen intuitively on varied micro and macro levels. Some of this literature focusses on genre-based inquiries into mixing for markets (Krotz & Hodgson, 2016) and Hip Hop (Shelvock, 2016). This literature is evidence that the academy values the tacit knowledge of current professionals and that there is a need for deeper understandings of modern creative practices because this information can inform pedagogical approaches.

Practice-based research in the field of mixing is also emerging and these types of studies support the idea that the music-making activity provides an integral part of the research prior to reporting, analysis and peer review. The special edition of the International Association for the Study of Popular Music (IASPM)'s journal in 2017 is an example of this. My article (see Anthony, 2017a) explored the evolving role of the third-party mixer by mixing an artist's song whilst also engaging in production decisions for the work, thereby blurring the lines between the mixer and producer. This is a process that is occurring regularly among third-party mixers in the professional world and the pre- and post-mix audio stands as a musical representation of the research.

There are varying contexts in which mixing is undertaken. In traditional models, the producer and band would record the musical work and then move into the mixing process as the final creative stage. The faders would be pulled down and the mixing board reset to a zero condition, and the song would be mixed from the ground up. Then, in the 1980s, seminal third-party mixing engineers like Bob Clearmountain entered the workforce. Producers would send their work to these people to cast a new perspective on the music during the mixing process. Finally, with more music being produced within the DAW and within tracker/top-liner models, mixing began to occur during the production process itself. Programmers would apply suitable side-chain compression, equalisation and effects to the music within the DAW as they composed and produced the track. In this workflow, the traditional "final mix" practice does not necessarily exist apart from some final (minimal) adjustments to reverbs, levels and equalisation, for example.

In my opinion, mixing is where the music production process ends. However, "the last creative step in the audio production process, the bridge between mixing and replication (or distribution)" is mastering (Katz, 2003, p. 12). Note the word *audio*, not music, in the Katz quote. I make the distinction here because contrary to what some mastering engineers believe, it is my opinion a track should go to mastering so the audio can be sweetened and prepared for the platform it is to be released on. There should not be any musical decisions made, such as changes to song structure or the addition of further musical elements, during the mastering process unless under the specific guidance of the producer. I grant that this is a controversial viewpoint, but these are the cool little arguments that inform the competitive banter between rivalling professionals. My viewpoint in no way undermines the importance of mastering. Indeed, it is integral to ensure the most appropriate equalisation, limiting, dynamic range, perceived volume and perhaps out of phase information for the platform on which the music is to be released. This can be as diverse as streaming services like Spotify and YouTube, on a CD or on the currently resurging vinyl format. These platforms all have specific delivery requirements and a good mastering engineer will provide the artist with a selection of files that are suitable for the various release platforms.

The question that remains is: should aspiring popular musicians learn how to mix and master? In my opinion, absolutely. The industry is changing, any musical or audio related activity a popular musician can do (and do well) helps them to become more in-control of their own activities. This also minimises expenditure in areas that are achievable by any person (to a modest level) with some training. The key here is to know when one should do these things and when it is important to get a third-party person involved for the betterment of the music. The pros and cons must be weighted, especially against the consideration of what an outside person would bring to the musical work that has not already been achieved. These are important concerns for the producer to contemplate as they work in the modern industry.

Session management and entrepreneurial workflows

A popular music producer is often akin to a project manager. PMP involves an enormous number of organised activities: meeting with clients, record companies and managers, negotiations of budget and royalties, bookings of studios, engineers, musicians and additional equipment, as several examples. Time management is also essential to the producer so that any deadlines can be met and budgetary restraints respected. Rehearsals and pre-production sessions need to be organised in suitable places and within a realistic time frame. A producer's role sometimes includes making wake-up calls, organising food and alcohol and maintaining suitable work/party hours (usually around 12–18 hours a day) so that the session does not spiral into a pit of exhausted, partied-out personnel. Also, sessions need to be organised: what needs to be accomplished and who is required? Personnel need to be managed and nurtured. For example, it is healthy if the drummer wants to go and play a game of golf on a day that they are not needed. If a singer wants to sing at 2am because that's when they feel the vibe, it must be done.

It is common for sessions to get bogged down and minimal work accomplished. In these scenarios, it is important that producers keep projects moving to the best of their ability. Bob Ezrin suggests that it is better to "do *something* rather than sit around scratching our heads not knowing *what* to do. Let's do something and react to it" (as cited in Burgess, 1997, p. 22). Producers are *doers* no matter the consequence. Mistakes are part of the process; the correct path is easiest to find after experiencing the wrong one. Producers are fearless and there is little room for negative reflection when mistakes occur. Professionals think "well that sucked, but thank god I worked that out so we can go this way!" It is important the leader of the production (artist, producer, songwriter) does just that: lead, organise and exemplify the practice they want from others. Work hard, and if suitable party harder – whatever is needed.

Most of the organisational processes I have listed here can be classified as autonomous entrepreneurial workflows. This is a term I like to use with my students, and it will be one I discuss throughout this book. These entrepreneurial workflows continue into realms of industry integration, career development and longevity. For example, prior to the invention of the Internet and the DAW, consumers of music purchased physical copies, and producers earned a percentage of sales of creative works, working within large format recording studios where they often employed sound engineers, musicians and associated personnel (Burgess, 2014). Historical PMP practices such as these were very specialised; producers were often recognised more within particular genres of music and most recordings also incorporated the involvement of a record company. Therefore, the practice of PMP was often from a genre-biased and (often) commercialised perspective (Burgess, 2008). However, the music industry has changed, and the practice of PMP has evolved to incorporate a broader scope of engagement.

Consumers of music now have the option to stream and/or download their preferred songs or singles, as compared to buying albums, and this has minimised the sales potential for which producers can earn royalties. Consequently, music producers have needed to diversify their practices to maximise their output, working out of a variety of recording environments including large facilities, home studios or even with a laptop in a bedroom (Burgess, 2008). These producers use the Internet to collaborate and are free to undertake productions from many diverse genres that are often self-released by the artist or the producer's own production company. To compensate for lost income, music producers have become entrepreneurs, a "one-person band" (Burgess, 2008), and engage with many roles including, but not limited to, engineer, performer, songwriter, programmer and record company executive. This is important in the modern music industry because being multidisciplinary diversifies income potential for music producers and this promotes career longevity (Anthony, 2017a).

SUMMARY

Chapter 2 and this chapter have targeted the many workflows involved within the greater practice of PMP. However, these elements can be nuanced by considering the music's genre or style (Pop, Folk, Country, Rap, for example) and the socio-cultural and geographical affordances of the people involved. The pedagogical frameworks that will follow in this book will therefore be adaptable enough to be suitable in many contexts and for students from diverse backgrounds and musical preferences.

REFERENCES

Anthony, B. (2017a). A final guiding hand: A practice-based study into the evolving role of third party mixers. *International Association for the Study of Popular Music, 7*(2), 41–47. DOI: 10.5429/2079-3871(2017)v7i2.7en

Anthony, B. (2017b). Mixing as a performance: Creative approaches to the popular music mix process. *Journal on the Art of Record Production, 11*. Retrieved from https://www.arpjournal.com/asarpwp/mixing-as-a-performance-creative-approaches-to-the-popular-music-mix-process/

Anthony, B. (2022). The places and spaces of popular music production pedagogy in higher education. In B. Powell & G. D. Smith (Eds.), *Places and purposes of popular music education: Perspectives from the field* (pp. 265–270). Bristol: Intellect.

Beato, R. (2018). What makes this song great? Ep.18 Max Martin. YouTube video. Retrieved from https://www.youtube.com/watch?v=cBDJDZAo41c&t=1052s

Bennett, J. (2018). Songwriting, digital audio workstations, and the Internet. In N. Donan (Ed.), *The Oxford handbook of creative process in music* (pp. 1–30). New York: Oxford.

Burgess, R. J. (1997). *The art of music production* (3rd ed.). London: Omnibus Press.

Burgess, R. J. (2008). Producer compensation: Challenges and options in the new music business. *Journal on the Art of Record Production,* (3). Retrieved from http://arpjournal.com/producer-compensation-challenges-and-options-in-the-new-music-business/

Burgess, R. J. (2014). *The history of music production.* New York: Oxford University Press.

Gearspace. (2020). Q+A with Jens Bogren-brought to you y Bogren digital. [Online]. Retrieved from https://gearspace.com/board/q-a-with-jens-bogren-brought-to-you-by-bogren-digital/

Grande, A., Martin, M., Kotech, S., Kronlund, A., & llya. (2016). Into you [Recorded by Ariana Grande] on *Dangerous Woman* [CD]. Stockholm, Sweden: Republic.

Harding, P. (2016). Top-down mixing: A 12-step mixing program. In R. Hepworth-Sawyer & J. Hodgson (Eds.), P*erspectives on music production: Mixing music* (pp. 62–76). New York: Routledge.

Hepworth-Sawyer, R., & Hodgson, J. (Eds). (2016). *Perspectives on music production: Mixing music.* New York: Routledge.

Howlett, M. (2009). *The record producer as nexus: Creative inspiration, technology and the recording industry* (Doctoral dissertation, University of Glamorgan). https://eprints.qut.edu.au/39093/2/c39093.pdf

Izhaki, R. (2008). *Mixing audio: Concepts, practices and tools.* Oxford: Focal Press.

Katz, B. (2003). *Mastering audio: The art and the science* (2nd ed.). Burlington, MA: Butterworth-Heinemann.

Kealy, E. (1979). From craft to art: "The case of sound mixers and popular music". *Sociology of Work and Occupations, 6*(1), 3–29. Retrieved from https://doi-org.libraryproxy.griffith.edu.au/10.1177/009392857961001

Kealy, E. (1982). Conventions and the production of the popular music aesthetics. *Journal of Popular Culture, 16*(2), 100–115. Retrieved from https://doi-org.libraryproxy.griffith.edu.au/10.1111/j.0022-3840.1982.00100.x

Krotz, A., & Hodgson, J. (2016). Mixing for markets. In R. Hepworth-Sawyer & J. Hodgson (Eds.), *Perspectives on music production: Mixing music* (pp. 140–152). New York: Routledge.

Lauzon, D. (2016). Mixing in pre-production. In R. Hepworth-Sawyer & J. Hodgson (Eds.), *Perspectives on music production: Mixing mus*ic (pp. 114–121). New York: Routledge.

Mixerman. (2010). *Zen and the art of mixing.* Milwaukee, WI: Hal Leonard Books.

Moylan, W. (2017). How to listen, what to hear. In R. Hepworth-Sawyer & J. Hodgson (Eds.), *Perspectives on music production: Mixing music* (pp. 24–52). New York: Routledge.

Paterson, J. (2016). Mixing in the box. In R. Hepworth-Sawyer & J. Hodgson (Eds.), *Perspectives on music production: Mixing music* (pp. 77–93). New York: Routledge.

Schmidt Horning, S. (2012). The sounds of space: Studio as instrument in the era of high fidelity. In S. Frith & S. Zagorski-Thomas (Eds.), *The art of record production: An introductory reader for a new academic field* (pp. 29–42). Burlington, VT: Ashgate.

Shelvock, M. (2016). Groove and the Grid: Mixing Contemporary Hip Hop. In R. Hepworth-Sawyer & J. Hodgson (Eds.), *Perspectives on music production: Mixing music* (pp. 188–197). New York: Routledge.

Toulson, R. (2021). *Drum sound and drum tuning: Bridging science and creativity.* London: Taylor & Francis.

Toynbee, J. (2000). *Making popular music: Musicians, creativity and institutions.* New York: Arnold.

Tuan, Y-F. (1979). *Space and place: The perspective of experience.* Minneapolis: University of Minnesota Press.

Walther-Hansen, M. (2020). *Making sense of recordings: How cognitive processing of recorded sound works.* New York: Oxford University Press.

Watson, A. (2015). *Cultural production in and beyond the recording studio.* London: Routledge.

Williams, A. (2011). *Unpublished paper* presented at the 2011 conference of the Society of Ethnomusicology, Philadelphia, PA.

4

Songwriting and production

The holistic importance of the song within the practice of PMP is represented by how it bares influence on not only the performance and production process but also the "projected media image of the person or persons performing the song" (McIntyre, 2001, p. 106). Bennett (2011) suggests:

> [T]he song is the script from which the rest of the popular music production chain reads; it leads to the recording session, the artist's vocal performance, the promotion and distribution of the final mixed audio artefact, and thereafter implicitly to tertiary activities – gigs, airplay and merchandising.
>
> (p. 5)

Within any modern definition, the term *song* now encapsulates not only "lyric and melody but also the other recognisable components [that constitute] aspects of a musical work, [including] form and structure, rhythm, harmony, arrangement, performance and production characteristics that enable [the manifestation of the work in] material form" (McIntyre, 2001, pp. 109–110). Subsequently, popular music producers should consider songwriting's role in (1) the production and performance process and (2) the domain of popular music songwriting itself.

Songwriting includes a broad skillset that ranges from writing lyrics and melody with some form of instrument, to using technology to develop backing tracks and/or beats that inspire lyrical and melodic form. Subsequently, Cope (2013) suggests that "[t]he manner in which the process of production is married to the composition is … extremely important" (p. 4). Producers must go further than considering what songwriting is, they must develop an understanding of how good practice in the songwriting/production relationship can positively influence their PMP activities. Elements to consider could include lyrical, musical and sonic methodologies, and music producers often design the recording process to promote the artistic intentions of the artist and the song's *function*. A song's function can be loosely categorised in two ways (1) to target the commercial market or (2) to express a domestic or personalised perspective or message. Both approaches consider a specific end listener.

DOI: 10.4324/9781003055099-6

LYRICAL MESSAGING AND THE SONG'S FUNCTION

Joe Bennett had some insightful comments about the lyrical content of songs that were informed through his discussions with songwriting colleagues:

> One of the questions I want students to answer through analysis is: what are the lyrics in pop songs about? If you ask most people who've never written a song before to write a song, they might write about something mundane or domestic, or perhaps something deeply personal to them, like a loss. A songwriting teaching colleague once said to me: "song lyrics are about three things: love, dancing or self-assertion". Why should this be? I suggest it's because those themes represent what we use pop music for, as a society and as listening individuals.
>
> (Joe Bennett interview, April-May, 2021)

When music is consumed in society it creates commerce. The commercial influence on PMP is significant, and many producers will study the current Billboard (or locally relevant) charts and exercise emulatory practices so their works are tailored to compete with that which is currently popular and consumed by the public. However, in my experience, many aspiring musicians still write music from a more personal and introspective perspective, something that is less constrained by the wants of consumerism. A song and its production, in this instance, are more inherent to the artist's subjective preferences and lived experience. I consider both of these approaches as valid, but I do believe that the producer needs to be explicitly aware of the function of the music, whether to express, to target the commercial market or both, so that this informs their practice. This is where the producer archetypes become useful; producers can position, and also adapt, their practice within musically dictated guidelines (craftsperson, innovator or auteur).

THE SONG/PRODUCTION RELATIONSHIP

The influence that songwriting has on the PMP process is constant. As such, the two are interconnected from ideation to manifestation. Joe Bennett suggests:

> If someone is writing a contemporary mainstream pop record now, writing the lyrics is just another production duty, along with programming the hi-hat or choosing the sample set. Songwriting and production are increasingly convergent disciplines, and I would argue that in the mainstream pop industry they have completely converged. When you're writing a vocal hook, it doesn't matter whether you sing it in a single take, whether you put it together as a number of samples, or whether you comp (composite) it by editing together many takes.
>
> (Joe Bennett interview, April-May, 2021)

In modern PMP practices, an artist does not always write a song to the completion stage and then record it unchanged. The song, its form and function are considered from the beginning to the end of the music-making process. This may start during the playing of demos or programming of backing tracks, continue during pre-production and recording and culminate in mixing and mastering. During mixing, producers may, for example, raise the volume of certain lyrical content to emphasise meaning or complete any edits to promote the perception by the listener that the song is a single or an album track.

In this situation, producers benefit if they are either songwriters or have an in-depth knowledge of songwriting. It is advantageous if producers understand the intricacies of melodic and chordal relationships, have a feel for song structure and can support the lyrical messaging and meaning of the song via sonic design. Producers can acquire this knowledge without being songwriters themselves. They can do this from within HE education songwriting courses or by working with, and learning from, artists in the professional world.

The songwriting and production relationship has the potential to be both flexible and/or fluid and this is primarily influenced by a song's function. This means that song/production agency changes in reaction to the music one is making. Cope (2013) describes the importance of the various "ingredients" of the song/production process:

> Some genres of music are typically more dependent on some ingredients above others. For example, folk songs tend to rely very heavily on lyrical content because they commonly exist primarily as a platform for the lyric to say something. There may be an age-old story that is being conveyed, supported subtly and sensitively through a simple chord progression and arrangement.
>
> (p. 9)

In contrast to this, music like Hip Hop and Rap may lay more focus on the "interplay of rhythmic elements in the production [and these techniques are utilised because the] 'vibe' of the track is often what the song is really all about" (Cope, 2013, p. 9). Currently, I do not believe that modern songwriters and producers consider genres as hard boundaries; genre is now, more so than ever before, representative of a collection of sounds, concepts and techniques that all music makers can fuse with other genre characteristics should they choose. Genre blurring practices are now prolific. However, I do believe it is most helpful within the process of production to understand the function of the song, what the song/production relationship is trying to portray to the listener and the ways one can achieve this.

MUSICAL AND SEMIOTIC CONSIDERATIONS WHEN PRODUCING SONGS

In previous chapters I have alluded to the importance of considering how the music connects or communicates to the end listener. This connection, which is solicited by the intentions of the artist, can manifest by way of an emotional response (see Cope, 2013, p. 11), a personal connection like an enticement or embodiment of movement like dancing (see Howlett, 2009) or a connection to a part of someone's life. Songwriting and the production process come together using lyrical, musical and sonic semiotic[1] approaches to aid these connections between the music and the listener.

Semiotic messaging is an approach in which meaning is communicated or construed in abstract ways. Iconic producer, T-Bone Burnett, alludes to this when discussing how sounds and songs connected to him. He reflects: "to me, there was just this dark, mysterious world that music could evoke, and that's all wrapped up in tone" (as cited in Massey, 2000, p. 27). Burnett is suggesting that music can use instrument tone (dark in this example) to evoke an emotional response within him that connotes a mysterious place. Music production can achieve this in many ways. For example, songs that target love or dancing can pursue a connection with the audience through appropriate grooves and feels. I am no musicologist but, as a producer, my critical listening informs this in many ways. For example: my wife Kylie (who is not a musician) loves the single "Smooth Operator" by Sade (Adu & St. John, 1984). Recently I played it to Kylie and asked her why she loves the song so much. Kylie said: "It's Sexy!... It's makes you want to move... [and] Sade's voice is smooth and flowing". The song's lyrics speak of the allure of a "diamond life ... lover boy ... and maximum joy" – the type of love and associated life that is tempting and alluring but often fraught with danger. "Smooth Operator" is a song about love and it makes Kylie want to move (dance). These are two of the topics that Joe Bennett suggests are commercial songwriting targets. However, there is much musical and semiotic detail that explains Kylie's reactions.

"Smooth Operator" generally consists of a repetitive i v iv chord progression (Dm7, Am7, Gm7) played on an electric piano pre-set that is accompanied by a drum and percussion groove. As a producer, I associate this repetitious musical foundation to an engagement with everyday life and the mundane. However, because of the minor key, a reflective mood is expressed. A wonderfully compressed bassline with a mid-range emphasis interacts with this rhythmic foundation. Occasionally, though, the bassline is slightly less constrained than the repetitive drum beat and percussive elements and it moves through the minor chord progressions with playful melodic freedom. The combination of tempo, groove, instrumentation, chord progression and sound selection establish a foundation that makes Kylie want to sway and move, as if slow dancing with someone. Indeed, Kylie embodied this when I played the song to her in our kitchen! Slow dancing elicits a connotation of two people, one-on-one, lyrically alluding to the person that Sade is singing about (the titular Smooth Operator).

Another important component is the saxophone, which no doubt helps Kylie connect nostalgically to her teens in the 1980s where the use of saxophone was profuse in popular music. The saxophone part is an improvisation spanning the entire song. For me, this represents the male character in the lyrics who lives life within the restraints of the chord progression (life), but at the same time lives a playful and less restrained existence. Sade's smooth and laid-back lyrical delivery immerses the listener in introspection and nostalgia, as if Sade is reminding herself of her experience of falling for the wrong guy. Sade's performance is emotive and many listeners can relate to these types of life experiences. As a result, they connect to the song. The music and real life are one for the listener – the song succeeds in its ability to merge the music's vivid narrative and real life for the listener.

In the above paragraph I have outlined how songwriting, performance and production aspects are interrelated in connecting to the end listener. I acknowledge that my analysis is subjective, but so is the production and consumption of music. Producers must trust their subjective instincts while bearing in mind that consumers will experience and connect to music in different ways. Therefore, producers should value the subjective and intersubjective interpretations of others to inform their work. Being aware of these considerations increases the possibility that listeners will engage with their music and potentially heighten commercial success. Music analysis like this is a common and well-researched practice, and a high-quality example of this is Joe Bennett's (2019) discussions on Billie Eilish's hit "Bad Guy" (Eilish, & O'Connell, 2019). Popular music producers can learn a lot from the semiotic analysis of various hits because techniques can be stored for repurposing during later productions.

Much of my own skillset here is derived from my own years of experience in PMP – my aural library. In the discussion above, I mentioned that I thought the saxophone represented the male described in the song. I learnt this approach to musical analysis through my work with legendary film composer, Hans Zimmer. I was fortunate enough to engineer and mix the soundtrack for the movie *Greencard* scored by Hans (Weir, 1991). Hans mentioned that he liked to talk with the actors about their characters and then develop melodies or choose instruments that are relatable to that actor's character. In the movie, Hans utilised the fretless electric bass to represent the character of George played by Gérard Depardieu. Subsequently, the bass was the lead instrument that played the musical theme for most of the music in the film; it was dominant in this regard, as was George's character. Similarly, in Sade's song, I associated the saxophone with the male character in the song's lyrics (and shown, also, in the promotional video clip). Consumers of popular music do this in their own nuanced way, influenced by their own social and cultural background, and their interpretations may or may not align or with the composer/lyricist's intentions. The relationship between songwriting and PMP practice facilitates all of this, and the proliferation of the DAW has only extended the possibilities of this symbiotic relationship because it integrates the production and songwriting process to a combined form of creative practice.

TECHNOLOGY AND SONGWRITING

I think it is important to cursorily explore the best-known research regarding the studio as a compositional tool. Moorefield (2005) highlights the compositional affordance of audio technology during creative practice. He suggests that producers are able to compose with recording technology because composition is interpreted more literally nowadays, meaning that technology facilitates the placing of musical items together and this results in composition happening. Brian Eno's concept of using the studio as a compositional tool (as cited in Cox & Warner, 2004) further exemplifies these notions. Eno's explanation suggests that the studio is similar to a musical instrument and therefore possesses compositional properties. *Oblique Music* (Albiez & Pattie, 2016) comprises a collection of chapters dedicated to investigating Eno's work as a musician and producer. The whole book infers that there are vast musical and emotional responsibilities that are interrelated with PMP, and the recording studio is one of many musical instruments producers can use to address these. Over the past twenty years, one piece of technology has arguably influenced the craft of songwriting to a greater extent than any other: the DAW.

Working with songwriting and the DAW

Marrington (2017) establishes a number of ways that the DAW mediates the songwriting process. He suggests that DAW's have their own properties which influence a songwriter's agency (p. 85). For example, the DAW's "'cut,' 'copy' and 'paste' ... [can] potentially turn the composition process into a graphic design exercise" (Marrington, 2017, p. 80) where a song's form is determined through the alignment of patterns. "Linerazation" is a term Marrington uses to discuss "the tendency of most DAWs to encourage the organization of material on a timeline, with varying degrees of flexibility for experimenting with alternative configurations" (ibid.). Subsequently, this promotes a compositional process where music can be "built additively by appending one item after another until the desired duration is achieved" (James Mooney as cited in Marrington, 2017, p. 80). DAWs also allow the songwriter to utilise "accumulative" compositional processes. This is where "the constituent parts of the music are gradually introduced one layer at a time until the piece appears as a complete entity in front of the user" (ibid., p. 80). These examples suggest that if songwriters develop an appreciation for the compositional possibilities of the DAW and, "perhaps more importantly, a sense of whether the tool might be pushed outside these boundaries" (ibid.), this is likely to have a positive effect on the resultant songwriting and production.

The versatility of the DAW as a songwriting/production tool is evident in Rihanna's hit "Umberella" (Harrell et al., 2007). When "Umberella" was released, it is intriguing to consider that the song was not written with Rihanna in mind. Seabrook (2015) writes:

> The song was the work of three men: Tricky Stewart, a producer, Terius Nash, a singer -songwriter who called himself "The-Dream"; and Kuk Harrell, a vocal producer. Together they make up the songwriting team at RedZone entertainment, a music production company and publishing firm based in Atlanta.
>
> (p. 187)

These people work in conjunction with a production company. They are not just songwriters, and when Kuk Harrell was questioned he suggested that their success came from approaching songwriting in a similar manner to working in the jingle business: "we were trained to do exactly what the client wants you to do, and do it when they want it" (Kuk Harrell as cited in Seabrook, 2015, p. 188). There is little consideration here for personal expressivity, these are workflows that explicitly target the Billboard Top 100. The description of how "Umbrella" was written highlights how the production and songwriting process was completely integrated. Seabrook unpacks Harrell's descriptions:

> 'I was fooling around with Logic' he says, trying to learn it. And I had gone into the samples and found this hi-hat loop, which I put on a beat'.... 'Then Tricky comes in and says, 'What's that?' Stewart sat down at a keyboard and started playing chords into the box, over the looped high hat and snare sound. Then he programmed a bass line. At that point Dream came into the studio, listened to the track and the word 'umbrella' popped into his head. He went into the vocal booth and got on the mike, singing 'Under my umbrella'. And then, inspired, added the all-important echoes – 'ella ella ella eh eh eh" – that became the song's signature hook.
>
> (Seabrook, 2015, pp. 188–189)

The song was originally intended for Britney Spears but was rejected before Rihanna picked it up and a world-wide hit followed. The producer's jingle-like workflow enhanced the song's suitability for the commercial market and any singer who could fit that market could take advantage of the song's effectiveness. In this example, the message behind the lyrics is stereotypical of Bennett's attestations above: two people under one umbrella equate to love. The singer's personalised self-expression (outside of love, dancing and self-assertion) was not the main influential factor of the writing process. Rather, this song's function was little more than to be consumed by the public.

There is fluidity and interconnectedness in contemporary pop songwriting. Thompson (2019) outlines this as "1) the structure and form of popular songs, 2) the musical elements of rhythm, melody and harmony, 3) lyrics and lyrical conventions, and 4) recording and music technologies" (Thompson, 2019, p. 118). In the manifestation of "Umbrella", interaction with the DAW (in this case the software, Logic Pro) was integral in bringing these elements together. The same song would not have resulted in its absence. As I have suggested earlier, the DAW is a mediating device because it enables varying methods and approaches. In the case of "Umbrella", it facilitated the writing and programming of a song where lyrics and melody were not formed through personal or emotional inspiration to begin with. In this instance, the songwriter was apparently learning a new software program and the song evolved from this process before collaborative engagements occurred, lyrics were written and the song was completed through inspired trial and error. "Umbrella" is a prime example of the integration of songwriting and production as a modern form of popular music practice.

SUMMARY

The overarching influence that the song has on PMP, and the fluid relationship between production and songwriting in commercial practice (as exemplified in the musical examples above), reinforces the idea that these two practices should be integrated in PMPP. Adam Bell's (2018) work, *Dawn of the DAW*, supports this theory by suggesting that the line between producer/songwriter and musician is ever eroding. It stands, then, that to facilitate PMPP without emphasising the relationship between production, songwriting, song meaning, function and context is limiting the student educational experience.

REFERENCES

Adu, S., & St. John, R. (1984). Smooth operator [Recorded by Sade] on *Diamond Life* [LP record]. London: Epic.

Albiez, S., & Pattie, D. (Eds.). (2016). *Brian Eno: Oblique music.* London: Bloomsbury.

Bell, A. P. (2018). *Dawn of the DAW: The studio as musical instrument.* New York: Oxford University Press.

Bennett, J. (2011). Collaborative songwriting: The ontology of negotiated creativity in popular music studio practice. *Journal on the Art of Record Production* (5). Retrieved from https://www.arpjournal.com/asarpwp/collaborative-songwriting-–-the-ontology-of-negotiated-creativity-in-popular-music-studio-practice/

Bennett, J. (2019). Anatomy of the Hit: Billie Eilish's "Bad Guy": Is she trying to subvert pop? Duh. Retrieved from https://tidal.com/magazine/article/billie-eilishs-bad-guy/1-55394

Cope, D, (2013). Song production: The marriage between composition and audio production. In R. Hepworth-Sawyer (Ed.), *From demo to delivery* (pp. 21–38). New York: Routledge. https://doi.org/10.4324/9780080928432-9

Cox, C., & Warner, D. (2004). *Audio culture: Readings in modern music.* New York: Continuum.

Eilish, B., & O'Connell, F. (2019). Bad guy [Recorded by Billie Eilish] on *When We All Fall Asleep, Where Do We Go?* [Album]. Los Angeles: Interscope.

Harrell, K., Nash, T., & Stewart, C. (2007). Umberella [Recorded by Rihanna] on *Good Girl Gone Bad* [Album]. California: Def Jam.

Howlett, M. (2009). *The record producer as nexus: Creative inspiration, technology and the recording industry* (Doctoral dissertation, University of Glamorgan). https://eprints.qut.edu.au/39093/2/c39093.pdf

Marrington, M. (2017). Composing with the digital audio workstation. In J. Williams & K. Williams (Eds.), *The singer songwriter handbook* (pp. 77–89). New York: Bloomsbury.

Massey, H. (2000). *Behind the glass: Top record producers tell how they craft the hits.* Berkeley, CA: Miller Freeman Books.

McIntyre, P. (2001). The domain of songwriting: Towards defining the term 'Song'. *Perfect Beat: The Pacific Journal of Research into Contemporary Music and Popular Culutre, 5*(3), 100–111. Retrieved from: https://journals-equinoxpub-com.libraryproxy.griffith.edu.au/PB/issue/archive

Moorefield, V. (2005). *The producer as composer: Shaping the sounds of popular music*. Cambridge, MA: MIT Press.
Seabrook, J. (2015). *The song machine: Inside the hit factory.* New York: Norton.
Thompson, P. (2019). *Creativity in the recording studio: Alternative takes*. Cham: Springer.
Weir, P. (Director). (1991). *Greencard* [Film]. Touchstone Pictures.

NOTE

1 Semiotics is the study of meaning-making and meaningful communication.

5

Music production subjectivity and creativity

The creation of authentic popular music challenges the producer to consider innovative territories in composition and production alongside that of past traditions (Thompson, 2019). These practices are often developed as tacit abilities as music makers gain experience. Subsequently, Bennett's (1980) work, *On Becoming a Rock Musician*, outlines clearly that rock music, as a form of popular music agency, is often *learned* rather than *taught*.

> Becoming a rock musician is not a process that is steeped in the history, theory, and pedagogy of prestigious academies; nor is it a learning experience which is guided by an informal tradition of teachers and teachings. Becoming a rock musician is not even a process of apprenticeship. In fact, rock music is *learned* to a much greater extent than it is ever *taught* by teachers.
>
> (Bennett, 1980, p. 3)

During this learning process, Bennett suggests that rock musicians develop a *musical consciousness*: an awareness of music making. Similar notions of consciousness have been conceptualised in the field of neuroscience where musicians' senses are stimulated during performance and an awareness of practice is formed (Levitin, 2006). These two sources confirm that an engagement with music is often a personalised process and, therefore, one could deduce that, although PMP may include diverse musical styles and frequently involve group contexts, the learning of PMP may be specific to the individual. Subsequently, a producer's consciousness, awareness, experience and knowledge inform their practice. Aligning with these concepts, the earlier portion of this book presented ontological and epistemological understandings of PMP that incorporate skills, creative practices and the importance of music production cultures. But how are a music producer's decisions processed and acted upon? And by what process are they informed?

DOI: 10.4324/9781003055099-7

THE SUBJECTIVITY GAP

The popular music producer chooses specific tools and approaches in reaction to the song presented to them. Producers utilise their aural library and respond by way of their musical history and the expanse of their knowledge. This suggests that PMP involves an element of subjective decision making. This notion has been explored in the work of Cumming (2000), who attests to the relevance of subjectivity in music. Cumming's work "positions subjectivity and selfhood in the gap between the listener and music [creation], in effect, closing it by arguing that 'musical signification' occurs at their intersection" (Moreno, 2005, p. 284). Cumming suggests that "for a performer to communicate music's expressive content he or she would require a specific kind of emotional understanding [of the music]" (Cumming, 2000, p. 8). By doing so, Cumming postulates that the "musician's own 'character' will be heard in the choices he or she makes, so it is that his or her own 'subjectivity' appears" (p. 9). However, Cumming warns of the precariousness of confronting the gap "between an 'interior' state [that contemplates an expressive work] and the expressive content heard by others" (p. 8). She asks how students can confront this gap if they do not know what the gap is. Cumming's work is positioned within Western European Classical Music, but these concepts are also useful in the field of PMP. Students' negotiation with the subjectivity gap that exists between music creation and the listening experience of the recorded product is an important consideration in PMPP.

CREATIVITY

In discussions regarding PMP, Hepworth-Sawyer and Golding (2010) suggest that "the producer's role becomes a balancing act between enabling creativity and encouraging productivity" (p. 188). The use of the term "creativity" in this instance is ambiguous and requires some clarification. Runco (2007) presents an extensive review of scholarly research on creativity that discusses disciplinary and topical avenues for creative thinking and practice. These include critical distinctions such as the relationship between creativity and problem-solving, and creativity as a form of intelligence.

Literature specifically considering creativity in music is reasonably ample (McIntyre, 2007, 2008). The seminal work of Csikszentmihalyi (1996) suggests that "a musician must learn the musical traditions, the notation system, the way instruments are played before she can think of writing a new song" (p. 8). These theories are associated with Csikszentmihalyi's (2014) systems model for creativity which is represented in McIntyre's (2021) reconfiguration in Figure 5.1.

The model in Figure 5.1 shows the three main elements that interact within a system of creativity. In the systems model, the three elements consist of a culture or a domain – thoughts, beliefs, a set of rules and the various technologies of the culture. The second element houses the agents of the domain who strive for new developments and intend to create novel outputs. But these can only be recognised as novel in the presence of the

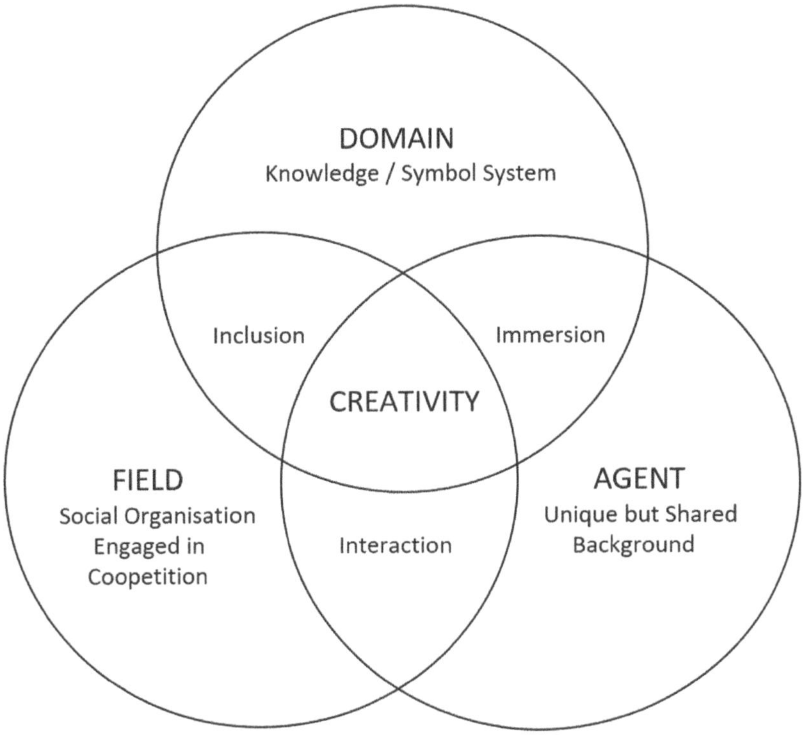

Figure 5.1 A reconfiguration of the systems model of creativity (Diagram reproduced from McIntyre, P. (2021) 'Songwriting Practice and Production: The Past, Present and Future', *The Songwriting Studies Journal,* 1(1), 5–6).

third element – the field. This systems model reacts in a cyclic fashion with one element informing the other and, as such, represents a foundational understanding for creativity in music.

Thompson and McIntyre (2013) have adapted these concepts and formalised an understanding of creativity within music production. They suggest that tradition (the many music production cultures, creative practices, and technologies that are available) can inform the producer's field of knowledge. This acts as a foundation upon which musicians can innovate to produce recordings of novelty as judged by the field: the public and fellow musicians. Considering creativity in this manner negates the common notions of it being intangible or unqualifiable.

Thompson (2019) is a more recent scholarly investigation into creativity in the recording studio. In this study, the author unpacks various domains including creativity in songwriting, studio performance, sound engineering and production. Thompson's recent work does much to elucidate the nature of inspiration in popular music. Thompson reflects that "prior to engaging in their respective creative areas, songwriters, studio performers, engineers and record producers [have] all acquired particular skills, knowledge, abilities and expertise" (p. 234). This knowledge was acquired through domain acquisition and formal and informal experiences, for example: working in the field, through immersion in diverse music production cultures and through listening and analysing music that has been made in the past (their aural library). The producer's repository of musical, technological and social information can be "employed and rearranged to contribute something new to the record-making process" (ibid., p. 234).

In light of Thompson's work, it is logical to surmise that PMP is reliant on the creative interplay of all the people involved in the music making. Romanticised notions that depict the artist "as the sole creative force inside the recording studio inadvertently diminishes the contributions of others (engineers, producers, and other musicians) and the necessary interactions with recording technologies" (Thompson and Lashua, 2016, p. 71). In these scenarios the makeup of the music production culture (the people, the technology, the music, etc.) has influence on the domain of knowledge that producers may draw upon.

It is interesting to consider the extent to which music producers are constrained by the song, its function or the genre of music within which it is situated. Paul Thompson suggests that:

> Every producer considers their audience to a greater or lesser extent – in Pop music, it's most overt. This ... simply means that they explicitly address the selection criteria of the field within their creative process. I think ... of these examples as synthesisers (just like all creative practitioners) – they bring different elements together in a unique way. The field then decides on the extent of that creative contribution.
>
> (Paul Thompson, email correspondence, May, 2021)

Thompson's suggestion that "every producer considers their audience to a greater or lesser extent" aligns with my previous discussions regarding producer archetypes (auteur, innovator and craftsperson). All of these archetypes operate within the context of PMP but attribute differing levels of importance to the expectations of their audience, from high (craftsperson) to low (auteur) degrees. The consideration underlining these outlooks is why, and by what influences, a producer would make a given decision. In the case of the auteur archetype, the foundation is largely personal and expressive in nature. However, to be deemed as useful, or even popular, all music must go through the validation process of the field upon release. The notion of popularity, however, is not a simple and universal metric. Teachers and students can have different expectations about what constitutes popularity. My colleague Nyssim Lefford discussed this:

> The democratisation of production tools, greater access to distribution and social media has created a situation where students do not necessarily need to understand how to appeal to a broad existing audience. Instead, they may expect to find a niche that connects them with an audience who values their specialised and subjective views. Again, these developments are not necessarily "bad" (or "good"), but they can cause misalignments between teachers and students.
>
> (Nyssim Lefford, questionnaire correspondence, May, 2021)

Prince Charles Alexander, however, advocates for a more commercialised consideration of practice where data points of popular production approaches are useful in creative music making:

> I don't know if you've ever been around any bebop players, but bebop players practice baba da bada bada dadaba do dobodobodomododidaba, right? And then in the solo, they're stringing line 1 together with line 346 and line 2 with line 212 and it sounds so fluid but they have practiced each one of those lines independently over a series of years. It's the same thing. I got these data points. I remember when I sat there for an hour and a half and figured out the gated snare sound [from the 1980s] … and so you remember all these data points and these come together in the composition. To me, that's what creativity could look like for the student. Now sometimes, people misconstrue what I'm saying, and they think that I'm asking everybody to sound like Finneas or Zedd, Right? Nooo!! That's not what I'm saying. What I'm saying is that there are data points that you gather from these guys and then when an interesting piece comes along in your composition, you say, "oh, I remember when Zedd did that. Lemme go listen to the Zedd record … ooooh! That's even better than what we're doin". So, now you have a choice. Get as good a Zedd or get better than Zedd.
>
> (Prince Charles Alexander interview, April-May, 2021)

Prince Charles Alexander's and Nyssim's words above resonate with Thompson and McIntyre's (2013) guidelines of PMP creativity. Data points and personal production approaches are all informed by PMP domain knowledge that can be called upon and appropriated, innovated upon or utilised for an extrinsic or intrinsic purpose. Producers can cherry pick at will from their aural libraries while they engage in creative practice. However, the creative process in both commercial and self-expressive works is always constrained in some way. Within the scope of commercial music production, Thompson (2019) identifies "traditions, practices, terminology, language and codes" (pp. 69–70) as elements that are important for producers to consider during the production process. If producers are focussed on self-expressive music, they are constrained by their knowledge and understanding of PMP. Subsequently, all music producers draw on "the symbols, codes and language [of music production, they] … rearrange and combine them in a unique way, in order to make something novel: in this case, a recording" (Thompson, 2019, p. 70).

CREATIVITY IN THE CLASSROOM

Scholarly work that investigates creative pedagogies has identified some important considerations for HE course design and content delivery. Cremin (2015) explores the challenges and possibilities of creative pedagogies for young children and asserts that creative pedagogies often involve "imaginative and innovative arrangement of curricula and teaching strategies" (p. 1). However, Cremin then suggests that within current research "a focus on arguably narrow notions of attainment dominates" (p. 1). Progressing chronologically to school contexts, Harris and de Bruin's (2018) case study of teachers' perceptions of classroom creativity suggested:

> [P]erhaps the biggest obstacle to developing creativity in our classrooms is not just allocating sufficient time, or appropriate and thoughtful testing, but subduing the fear of failing – a mindset prevalent in students, teachers, school administrators, and policy shapers.
> (p. 15)

These types of studies outline some of the challenges that may also be prevalent in a HE setting. However, the interconnectedness of PMP and creativity within practice-based applications signals clearly that these concepts must be thoroughly fleshed out, understood and experienced by students so that tacit abilities are developed.

A survey asking PMP students if creativity was clearly explained and engaged with in their HE courses received varied responses. Approximately 66% of students suggested that it was either not done at all, or done either minimally or in a confusing way. Some explanations of this response included: "creativity was just explained as what you like"; "From what I understand about creativity, everything that we create can be creative as long as it sounds good" (International student survey response, May, 2021). These answers show little knowledge of the domain-based theory.

Some students suggested they were left to their own devices to discover what creativity was:

> The 'creative' side was less explained allowing us (students) to find our own creative path for music production.... If someone has not defined what creative way to go then that person would be stuck in the same place without any help.
> (International student survey response, May, 2021)

Other students' experiences suggested that their curriculum focussed on creativity within the application of technology: "I don't think creativity was ever in focus since the education was based on technology, science and practice. Music production seemed more like a craft than an art, which gave me objective tools to make my creations more clearly" (International student survey response, May, 2021).

Even though I have utilised it in my classes, and I am sure others have as well, not one student discussed the systems model (Thompson and McIntyre, 2013; Thompson, 2019). However, some (about 5%) had a reasonable understanding of PMP creativity:

> This concept was described in detail, although perhaps not under a clear definition of 'creativity'. Over the course of my classes, we were given a detailed history of music and recording technology, and how these new developments affected music production. We also learned about the various skills necessary to effectively engage in music

> production (having a vision, communication/psychological skills, critical listening, musicianship, engineering, songwriting and arranging skills, etc.). Creativity in music production was therefore linked to the producer being aware of all of these factors and using them when required, utilising knowledge of past practices to approach each project as a unique entity and tailoring the production and chosen technology to best suit the artistic vision for the project.
>
> (International student survey response, May, 2021)

SUMMARY

PMP domain acquisition and an understanding of the systems model of creativity presents a platform from which HE students can engage with creativity and the subjectivity gap from an informed perspective. I would also postulate that PMP courses that eschew these types of approaches regarding creativity are omitting a vehicle that can empower students to convert cognitive visions into high-quality musical works. Therefore, the student's engagement with their own subjective process and the validation of their actions by the field are essential in the development of well-rounded professionals.

REFERENCES

Bennett, H. S. (1980). *On becoming a rock musician.* Amherst: University of Massachusetts Press.

Cremin, T. (2015). Creativity and creative pedagogies: Exploring challenges, possibilities and potential. *Education, 43*(4), 353–359. Retrieved from https://doi.org/10.1080/03004279.2015.1020632

Csikszentmihalyi, M. (1996). *Creativity: The psychology of discovery and invention.* New York: Harper Perrennial.

Csikszentmihalyi, M. (2014). The systems model of creativity and its applications. In D. K. Simonton (Ed.), *Wiley's book of genius* (pp. 533–545). Chichester: John Wiley & Sons.

Cumming, N. (2000). *The sonic self: Musical subjectivity and signification.* Bloomington: Indiana University Press.

Harris, A., & de Bruin, L. (2018). An international study of creative pedagogies in practice in secondary schools: Toward a creative ecology. *Journal of Curriculum and Pedagogy*. doi:10.1080/15505170.2018.1457999

Hepworth-Sawyer, R., & Golding, C. (2010). *What is music production? A producer's guide: The role, the people, the process*. Burlington, MA: Taylor & Francis.

Levitin, D. (2006). *This is your brain on music: The science of a human obsession.* New York: Penguin.

McIntyre, P. (2007). Learning to be songwriters: Creativity, the systems model and domain acquisition. In I. Collinson & M. Evans (Eds.), *Sounds and Selves: Proceedings from the 2005 International Association for the Study of Popular Music Australia/New Zealand Conference*, Ryde, Australia (pp.

46–57). Retrieved from https://search.informit.com.au/documentSummary;dn=171958987165814;res=IELNZC

McIntyre, P. (2008). Creativity and cultural production: A study of contemporary western popular music songwriting. *Creativity Research Journal, 20*(1), 40–52. DOI: 10.1080/10400410701841898

McIntyre, P. (2021). Songwriting practice and production: The past, present and future. *The Songwriting Studies Journal*, 1(1), 5–26.

Moreno, J. (2005). Review essay: Naomi Cumming, The "Sonic Self": Musical Subjectivity and Signification', Indiana University Press. *Music Theory Spectrum, 27*(2), 283–307.

Runco, M. (2007). *Creativity: Theories and themes: Research, development, and practice*. Boston, MA: Elsevier Academic Press.

Thompson, P. (2019). *Creativity in the recording studio: Alternative takes*. Cham: Springer.

Thompson, P., & Lashua, B. (2016). Producing music, producing myth? Creativity in recording studios. *International Association for the Study of Popular Music, 6*(6), 70–90.

Thompson, P., & McIntyre, P. (2013). Rethinking creative practice in record production and studio recording education: Addressing the field. *Journal on the Art of Record Production*, (8). Retrieved from http://arpjournal.com/rethinking-creative-practice-in-record-production-and-studio-recording-education-addressing-the-field/

Part 2

Popular music production pedagogy that bridges to the real world

This section of the book contemplates the integral link between the real world and higher education learning frameworks. Furthermore, a host of PMPP scholarly literature will be explored, which will be drawn upon in later chapters.

DOI: 10.4324/9781003055099-8

6

What is the real world?

I was writing an article in 2019 and, when the peer review came back, one of the revisions I was asked to make was to remove the term "the real world". In Australia, the use of this term is common. Students and educators use it in pedagogical conversations and institutions promote their connection to the real world in their advertising campaigns. In this context, the real world could be defined as what you will experience in a professional capacity outside of institutional training. I can understand the reasoning behind the review suggesting the term is problematic. Indeed, it may also bear a different meaning: university is not real. Of course, experiences in educational institutions are real, unless you and I are living in the *Matrix* and should take the blue pill!

For the purpose of this book, the term "the real world" is taken to mean the occurrences, both present and future, once a student leaves HE and enters the professional sphere. If one of HE's main aspirations is to prepare graduates for integration into the industry (as outlined earlier) then bridging HE experiences to the real world is an integral part of student development. However, it is interesting to note that many students do not consider university life as equal to, or the same as, the real world.

> No matter … how great and how creative and how even fun and engaging class is, it is still [just] class.… For us it's just being at uni again.
>
> (Hayley, EdD interview 1, February 26–27, 2018)

> Well, what comes to mind is the difference between the real world and uni. So uni is, uni unfortunately.
>
> (Georgio, EdD interview 1, February 26–27, 2018)

Within the specific context of the courses I have delivered, many students believed that the classes (lecture, tutorials, sessions) that they undertook were not like the real world because they happened at a university. As a result, my students (to begin with) lacked motivation, were poor at preparing for classes and often did not rehearse for performances or were not

DOI: 10.4324/9781003055099-9

prepared with session management strategies (lyric sheets or technical setup, for example). Part of being an educator includes an understanding that students do not magically embody the work of a professional practitioner. My approach in these situations is to present an authentic and honest opinion because I want students to feel like I am being genuine in my discussions with them. This helps to develop trust and enables opportunities for important, if often difficult, discussions to occur. In one of my classes, I probed students as to why they had not rehearsed for a tutorial session (see Video 1 on the website at www.routledge.com/9780367517632). (Please note some of the videos from my doctoral studies have subtitle text in them, these were utilised to identify themes in my thesis.)

> Hayley: I guess, like, the difference between organising our own stuff and organising our class stuff, there's a big difference...
> Lennon: Um, well, we have a lot of other stuff to balance as well that we're already rehearsing for or working on so finding... time.
> (Video 1)

Both Hayley and Lennon are devaluing their class activities as compared to their autonomous work and the resultant dichotomy that we face as educators, between the real world and university life, is ever present in my response to them.

> It's just interesting because it does happen a lot in class where students don't rehearse a lot for a class gig, but they whinge when it's not real-world learning. Do you hear what I am saying? You got to rehearse!

Perhaps in this example student *accountability* and *headspace* deserves consideration, but many of these problems manifest from students' negative perceptions of the HE experience. This is not an isolated example, as was discovered in my interviews with international colleagues including Lachlan Goold.

BRENDAN: Does uni come with a stigma: "oh, it's just uni?"

LACHLAN: Well, I try, as much as possible, for it to be more than just a simulation of the real world, but … it is difficult in a university environment. I think you're right … the majority of students go to university and treat it like university.

BRENDAN: What are the elements which are giving uni that stigma?

LACHLAN: You know, that's a very interesting question and I'd love to know the answer. Like, is it perhaps just the framework of university in that there's the constant assessments? Also, university doesn't really replicate the industry that well. You know, I had one band that came from Tasmania, and when you're not in the studio [working] – there's no pool room, there's no ping pong table …

BRENDAN: There's no X-Box …

LACHLAN: Yeah … it's a very formal environment and that [social element] really is part of the whole recording [industry].

BRENDAN: What you're talking about is the culture of recording.... The musical, technological and the social culture of making music.

LACHLAN: Yes. Very well put. And, a university can't really replicate that. But ... what it can replicate, is the "scene" because you're putting a lot of people from all different worlds, that don't know each other together ... and by the end of the three-year program, [through collaboration] they start to work out how to network ... (Lachlan Goold interview, April-May, 2021)

Similarly, Adam Bell discussed notions of a disconnect between the real world and assessment items:

> I think that [stigma] is sort of understood in a way, especially when it comes to popular music.... For popular music or any kind of assignment that I would give, I think most of the times it would be viewed as an assignment unless I can really relate it to what they're about to do in the real world.
>
> (Adam Bell interview, April-May, 2021)

Various international students' comments confirm a disconnect between university life and the real world.

> I felt disconnected to the industry during school because it all felt very hypothetical and with extreme attention to detail and being correct.... My experience of working [making music] these past years is that things are more roughly done, faster and more ad hoc, than in school.
>
> Since most of our studies were theoretical, I didn't really feel like some stuff was even applicable.
>
> (International students' survey responses, May, 2021)

However, students reflected positively when they felt they were working on a *real* project:

> It felt real when I made music & sound effects for the video game company, when I "worked" for them, as well as my own project phases when I "worked" for myself.
>
> (Ibid.)

THE REAL WORLD OF TODAY, TOMORROW AND THE FUTURE

If class activities can bridge to the profession, it helps the educational experience to feel real for students and promotes the learning of advanced concepts. For example, my research has shown that students benefit when they are in control of session activities, can enable their agency and react to their subjective preferences through intuitive decision making (Anthony, 2019). This would contrast with receiving, and following, instructions from the educator. There are many possible trial and error scenarios that could be useful here, but one example of a tutorial which, in my experience, garnered positive student feedback with respect to its real-world feel, was the running of a session that employed a live synth player. This type of session added complexity to the activity because it is common for the musician to try different ideas and require guidance regarding the track's needs. In the student feedback session, it was discussed that there did not seem to be any fear to create, and the students stated that they were excited and having fun and that "it wasn't like a class, it was like a session" (see Video 2 on the website at www.routledge.com/9780367517632).

During that video, you will notice I used some expletives in my commentary to students. This tutorial occurred many weeks into the course and, by this time, both the students and I were comfortable with that kind of lexicon in the class environment. It was utilised in a non-offensive way and was helping to balance out the educator/student dynamic. Students reflected positively on this tutorial:

GEORGIO: See ... I've never done a session like that,

HAYLEY: Yeah, I don't know what it was about it but just the way that we just got in there and just did stuff …

LENNON: I think it was cool because it was so trial and error.

GEORGIO: And it was self-directed too. It was just like us chatting, throwing around things, do we need this?... In that session I felt like I was taken seriously ... and in a way it was all like … that cross-pollination of influence as the innovator [archetype] ... that's what I hope audiences tap into, like "ooh, this is different"… so … I like that, yeah.

(EdD Focus Group 1, May 29, 2018)

The above discussion not only alludes to the pedagogical literature discussed in the following chapter (Kolb, 1984; Green, 2002; Lebler, 2006), but words and phrases like "innovative" and "trial and error" resonate strongly with professional practice and are important to highlight as pedagogical targets for PMPP.

Our job as educators is made challenging because PMP practice has changed rapidly over the last two decades and currently occurs in many forms: producer-led; co-produced; producer and band and self-produced. Jo Lord discusses this: "Often I find that facilities and course content are the biggest hindrance in future development of practice as they are often not updated regularly enough to reflect the fast-paced changes of the industry" (Questionnaire response, April-May, 2021). As portfolio musicians, students will benefit from experience in all the scenarios listed above. They will similarly be benefitted by having a broad selection of skills and knowledge in acoustic, electronic and online music making. Electronic music producers will benefit from instruction in working with singers and vocal performance, how to use microphones, how to tailor instruments' sounds and how to capture sounds (even if just for samples), as several examples. Acoustic music makers need to experience the compositional affordances of the DAW to promote the diversification of their work. All students will also benefit if they have knowledge of some of the more traditional approaches to PMP.

Traditional production approaches

Even though I am in support of targeting modern music production contexts within PMPP, I do think knowledge of traditional approaches has utility in expanding an aspiring producer or musician's aural library. I discussed these ideas with international colleagues.

> I think there's a place for everything and I think we should be reflecting the realities of sound recording practice as they are out there in the real world [modern PMP]. I think we should be giving students opportunities to encounter certain technologies and certain processes and practices that are the best of the historically informed stuff but also that give them perspective on how records have been made in the past.
>
> (Samantha Bennett interview, April-May, 2021, 2021)

> [Students] are often quite ignorant these days of what you might call classic production that relates to a lot of the music they'll still know of course … in terms of a legacy catalogue. But they will have little knowledge of how it was done. So, there's value in teaching music production that historically situates the practice.
>
> (Mark Marrington interview, April-May, 2021)

Students also value instruction in traditional production techniques:

> I love the sound of 60s and 70s records. I love the techniques, the vibe, the place that it takes you to.... All types of techniques past or present need to be known because there is no knowing what sort of clients you are going to come across.
>
> (Georgio, EdD interview 1, February 26–27, 2018)

These kinds of approaches include live recording creative practices (Anthony, 2015) which demonstrate the importance of generating high-quality musical performances. These activities also develop transferable skills including communication, psychology and critical listening and reflection. Some authors view these historical practices as outdated and inclusive of "insidious by-product[s]" in that they only speak to certain genres of music (Luttrell et al., 2020, p. 30). In some recent literature regarding the *MacGyverying* of the small studio (Luttrell et al., 2020) the authors have attested that live recording requires the affordance of "expensive industrialised recording spaces" (p. 30) and therefore are not an inclusive process for all popular musicians. I disagree with this opinion. Live recordings can be completed with only a handful of microphones and they can occur in a home studio or on location. Moreover, this observation spans genre. For example, Hip Hop live recordings (see Songkick, 2013) promote artists developing new versions of their material and this pedagogical application places emphasis on socio-musical processes as opposed to the use of tools or technology. Amandine Pras agrees:

> I like to teach a certain part of popular music production [not all of it] with musicians on site who are playing live together in the studio and so for that, you don't need an analogue console, but you need some kind of space.... I really like to show my students what it is like to compose in the studio with a group of people and not just the artist/producer type thing.
>
> (Amandine Pras interview, April-May, 2021)

Traditional approaches like live recording are versatile pedagogical tools but, in contrast, Luttrell et al. (2020) rightly point out the utility of also learning to be flexible within small studio and DAW-based production contexts. By utilising acoustic, electronic and hybrid methodologies, institutions provide students with the tools to be flexible in their future practice so they can autonomously specialise their craft in any direction that they feel suitable. Therefore, institutions should provide the studio affordances that students require for their musical interests and the settings they will find in the industry after their time in HE.

Large, small and home studios

It is no secret that many institutions advertise their expensive large-scale studios to attract students to their institutions and programs. Consumerism is rampant within the HE sector and this is an unavoidable part of the system under which we operate. However, I am an advocate for the realignment of HE facilities used in PMPP to be more akin to the studios that are active within the profession. This would most definitely include numerous small post-production rooms[1] and single-room studios that resemble a home facility. Goold (2018) found that the difference between recordings completed in large facilities as compared to home studios are indiscernible to the listener. However, it is also important to consider what large studios may bring as pedagogical tools for education. Amandine Pras suggested:

> In an ideal world, I think it's still worth it to have access to a large-scale studio.... There are costs associated with them ... but from my experiences ... students who get trained on [large consoles] tend to develop critical listening skills quicker ... and will adapt very well to reproducing this in the DAW.
>
> (Amandine Pras interview, April-May, 2021)

Amandine went on to suggest that large-scale studios also allow students to work in a social environment with which they are unfamiliar, and this helps to develop their social skills. Sam Bennett agreed and, in our discussion, alluded to the advantages of targeting students' communication skills:

> It's life skills, it's communication skills. I mean, anyone knows when you're recording, it's all about psychology and it's all about communication and that can be the difference between getting brilliant takes out of somebody in a recording situation or making them just walk out and never ever set foot in a recording studio again. And this is the

> problem with so much focus on the gear and so much focus on "my preamps better than your preamp, this microphone's better than that microphone" ... if you haven't got the communication skills, forget it.
>
> (Samantha Bennett interview, April-May, 2021)

I also discussed the place of large studios in HE with Mark Marrington

MARK: The small set-up is sort of increasingly becoming the norm and so obviously teaching has to adapt ...

BRENDAN: Do you feel there's still utility in those larger studios?

MARK: Well ... DAW-type software to a large extent is modelled on [Large studios] ... they're software recreations of these earlier historical production paradigms. So, to get [students] to engage physically in those environments, brings that into kind of a three-dimensional reality for them. I think it's that that gives them a deeper understanding.... I guess while DAWs still continue to refer backwards to things, then one still needs to think in terms of those older things as much as the new things.

(Mark Marrington interview, April-May, 2021)

Russ Hepworth-Sawyer had an interesting perspective on this:

> I've always thought they [large consoles] were brilliant, just for teaching purposes, they're just so clear and easy. . . . I have to admit, from the minute we bought a console ... the students understanding of basic audio theory, technology [has] just gone up and the music they've created has gone up [in quality]. [Also] we've only just got an analogue studio because we realised through market research that when students [are searching for a course], they want to see something they will never buy and they will never have and never probably use and that is a shop window for us.
>
> (Russ Hepworth-Sawyer interview, April-May, 2021)

Samantha Bennett agreed with this perspective suggesting that "pedagogical facilities have a really important role to play in providing students with both the technologies and techniques and access to facilities that they won't encounter at home" because,

> If you were to come to uni and all they had was a room full of laptops with Ableton on them (and I'm not saying that you couldn't) but ... what are you offering the students that they couldn't get at home when YouTube is full of videos on how to operate a DAW [and other production processes] ... I think we are duty-bound in lots of ways, to offer a range of opportunities that you can't get just from doing that.
>
> (Samantha Bennett interview, April-May, 2021)

Paul Thompson also added an innovative thought that some institutions are implementing:

> I see a time when potentially universities are going to become part of the recording industry, where they're going to loan out these spaces

> because they're the ones with the money to have larger facilities that can be maintained and run in a way that large facilities require. I think that could be a really cool thing, it could be the next step to apprenticeships or even just a next step to integrated learning and education.
>
> (Paul Thompson interview, April-May, 2021)

Conversely, Andy Edelstein argued that the money saved by not implementing large studio equipment could be put to better use by investing in "different resources, namely faculty support, course development, and research". Andy also likened the debate regarding large or small studios to doing "music production using flathead screwdrivers and music production using Philips head screwdrivers". This reaffirmed Andy's opinion that the development of a student's creative application and flexibility of practice, regardless of technology, is key (Andy Edelstein interview, April-May, 2021). A statement by Andy on the Berklee website reads:

> State-of-the-art recording studios, though quite nice, are not available to recording artists as often as in the past. But it can be motivating, even freeing, to divorce yourself from traditional practices. Lacking all of those high-end options, perhaps you'll think more creatively. You might come up with solutions you didn't consider before, like interesting acoustics or a different vibe, and that in turn will affect the way you go about the project – often in a highly positive manner.
>
> (Berklee, 2021)

Many students suggested that universities should include small home studios and electronic producer environments as instructional facilities (EdD interview 2, April 24, 2018; International student survey response, May, 2021). Feedback such as this was common:

> Stop living in the past and leave the Neve alone. Large format recording sessions are dying because no one does it like that anymore unless you're an elite in the industry. Help create traction for the students that are starting from literally nothing so they can become sharp with their product. When I say that, I mean it's nice to learn to track a band, and it's nice to have millions of dollars of equipment at your fingertips. But what else is nice, is using this thing called a computer (because most of us have one).
>
> (International student survey response, May, 2021)

Small/home studios are normally single-room environments (there is no separate control room connected to a recording room). If such a space is utilised, educational applications can demonstrate how to work within a small single-room space. Lachlan Goold highlighted the importance of the home studio in HE facilities:

BRENDAN: Is there a place for those home studios

LACHLAN: There has to be. So, like ... I was going from my home studio and one day a week I'd teach sessionally and I was walking into

> [the university studio] and I'm just like, "hey guys, this isn't the real world!" [Large studios are] becoming rarer and rarer, nothing's changing.... Perhaps I'm a socialist producer. I'm very much about teaching for the majority of people.... I think you definitely have to teach for the home studio.
>
> (Lachlan Goold interview, April-May, 2021)

Personally, I would prefer to have both large and small studios available in HE. Samantha Bennett makes a strong point (above) when she suggests institutions are duty bound to offer a range of technological opportunities that include large-scale studios, but I am also a bit of a socialist producer and agree very much with Lachlan's comments. Lachlan advocates for the inclusion of small studios based around UAD systems and the Console App. However, if institutions wish to include the types of facilities students will use at home, further consideration is necessary. Sam Vallen discussed this with me:

> UAD are a great company and their plugins and, increasingly, interfaces, are industry standard. The issue is that, firstly, their gear is expensive, and the cost of entry is probably far higher than what students can afford, which subsequently means students will get less for their money. The second issue is that the idea of having plugins tied to DSP feels decreasingly relevant in an industry where even mid-level computers can run hundreds of elaborate plugins without breaking a sweat. If all the students need are the plugins, brainworx have made all of their UAD stuff available natively on Plugin Alliance, and many other companies offer equivalent plugins, some of which can run natively OR take advantage of available DSP processing. For students entering the DAW-based recording environment, I'd sooner recommend contemporary companies like Valhalla, Slate, or Fabfilter whose gear is more affordable, and in many ways superior. Pair that with an entry-level interface and there's really nothing standing in their way.
>
> (Sam Vallen, personal correspondence, December, 2021)

I think it is important for matters of equity, inclusiveness and diversity that educators consider Sam's comments. Students are incredibly talented at finding a way to do things on a tight budget. Therefore, our duty (in my opinion) is to provide them with facilities that normalise the studios they can afford and build themselves. Institutions can also (should they choose) provide opportunities with expensive technology and within large-scale studios that graduates may encounter through their engagement with the industry.

When I spoke to Joe Bennett, he provided a different and program-specific consideration:

> I do think that the key to this whole debate is repertoire. What kind of music do you want the students to make, or do they want to make?...

> Most University degree courses market themselves on the sometimes true assertion that we are preparing you for professional activities as a musician; that's the overarching promise. But how are we defining that word 'professional'? I sometimes take the shortcut of defining professional partly by the centre of the mainstream of pop music – representing the 'top' of the profession, at least in terms of the amount of listener engagement. And if creating mainstream pop is an aspiration for at least some students, then we should ask ourselves how necessary it is to teach students in glass-wall studio environments. If we take a slice of the top 50 in the U.S.A. at any one time, only a tiny minority of those recordings required the micing of an ensemble in a glass wall environment. [So what are our programs asking students to do?] If we want our students to make EDM then we maybe don't need any kind of studio at all.... It would be feasible – desirable even – to teach an EDM course entirely in the box (in software) with maybe a couple of good mastering rooms.
>
> (Joe Bennett interview, April-May, 2021)

Joe's central implication is that HE institutions should match the purpose of their students' study with the kinds of facilities and resources found in their respective musical styles. If institutions aspire to develop industry professionals, then the institution must provide education regarding the modern music industry.

THE INDUSTRY

Russ Hepworth-Sawyer outlined to me the importance of educating students about the modern music industry and how to derive an income from it:

> The first thing that I introduce them to [students], is that the music industry as "they know it", is over. So, we show them the world isn't about this "I'm going to be a rock star sitting in a tour bus for the rest of my life". It's over! Move on! That world certainly doesn't exist at the moment.... The students then say well if this guy is saying the music industry is over, what the fuck am I doing here?
>
> (Russ Hepworth-Sawyer interview, April-May, 2021)

The way the general public currently consumes music has drastically changed the music industry from pre-Internet days.

> A business model for Spotify is now solidified, the agreements that the record labels have made with Spotify means that they get a solid physical payment that doesn't vary for their catalogue and the record companies are really happy because their cash flow's sorted, it's too easy for them. So someone's got to come in and disrupt that model.
>
> (Russ Hepworth-Sawyer interview, April-May, 2021)

If students acquire a modern understanding of how music is consumed, then their practice can negotiate with this. That does not mean they must tailor their activities to accommodate the current industry. I believe that musicians should be paid fairly for their work and not be exploited by corporate practices or various governments' inequitable decisions, an opinion that seems increasingly common. Therefore, educators not only need to help students learn the greater practice of PMP, they also need to empower students to acquire the necessary industry knowledge so they can forge an innovative future for themselves in a fast-changing world. Kirsten Hermes discussed this:

> There are no traditional job roles anymore, so you need to have a business mind to figure out how you can find work, and how can you target the people who might want to pay you for your skills.... This often includes developing a strong artist identity, along with a sonic signature, to develop a unique selling point.... So, I think that by reframing what skills actually are required, you can bridge barriers to entry in the job world.
>
> (Kirsten Hermes interview, April-May, 2021)

Kirsten is alluding to entrepreneurial and business skills that also cross over into project management applications. These are the tools graduates require to not only integrate into, but also innovate upon, the industry and potentially improve the working conditions of all musicians into the future. Zack Moir elaborated on the dangers of pedagogical approaches that do not support this:

> I think it is absolutely the potential for domestication. Thinking about that in a specific critical pedagogy sense, if we have an education system that is primarily based on the notion of social reproduction, then that applies to cultural reproduction. When that happens, what we're doing is we're retrospective in applying things that people have done to create certain things, so that we can teach people to do those things, such that the reproductive cycle keeps going. Actually, if that keeps going, and it has done beautifully if you think about something like classical music pedagogy for three hundred years, which is why you still get students sitting down to write a Bach chorale. Why the hell do they [students] want to sit down and write a Bach chorale!? Well that's how the pedagogy works.... I think that social reproduction not only glorifies particular methods of doing things, which therefore have particular results attached to them, but also that the bigger systems that sit around that, like the economy, like the silos of a producer, like the fact that a sound engineer is classed as a lesser link in the chain than a producer. That a producer sits above all of these things and ultimately makes the decisions. I think if we instil these things in the pedagogy, we're normalising them for the future and then we're failing to provide the opportunity for liberation.
>
> (Zack Moir interview, April-May, 2021)

THE REAL WORLD: DRUGS, ALCOHOL AND GETTING IT DONE

Anyone who knows me well knows I am a reformed alcoholic of 15 year's sobriety. Any addict will tell you there is something very powerful in someone who deals with their addiction every moment of every day, admits to the power it has over them unconditionally and, as a result, can abstain. This (in my opinion) is probably one of the most inspiring things a person can achieve in their life; a reformed addict is a remarkable person indeed. Right now you may be thinking "hang on, is this an AA book, or is it a book on PMPP?" Well, right now it's a little of both. However, if institutions aspire to educate students about the real world then there must be some honest and authentic discussions about the use of alcohol, drugs and the possibility of substance abuse.

I often discuss my experiences with alcoholism with my students if the appropriate opportunity arises, and I find it helps to facilitate an educational and enriching experience for the student. My message with regards to alcohol and drugs is that I do not condone any illegal processes, but I will not ignore that these activities occur in the industry with which we work. Therefore, I tell students my general rule is that once anything you are doing is affecting your work, then it's not on. For me, in a professional context, this is a good rule, but in a HE setting there is much more to consider. Many of the great producers I worked with never drank or took drugs while they worked, but many bands did.

When I wrote and recorded my own music, I would always drink. Over time this became such a regular and mundane part of my process that it got out of control. I never did any kind of drugs, but I could roll one hell of a joint if the client needed me too. It's a slippery slope. I often tell a true story to my students of an album I worked on with a guy called Kenny (I wrote some songs and played keyboards on the record). Kenny was hired on this session not as an engineer, producer or a session musician. No, Kenny had one all-important role: he was *the vibe.* Kenny's gig was to keep the fridge full of beers, get the lunches, roll the spliffs and tell the jokes. For that session, Kenny's role was deemed as important. The musicians were not getting paid that well, we were working long hours and needed to enjoy the session or else our performances would have suffered. Kenny was a valued member of the music-making team because he did much to nurture the vibe and help manifest the necessary music production culture to complete (what turned out to be) a highly successful album.

I have taken the liberty to pen some of these stories here. I do this because it replicates the way I communicate to students regarding the real world of PMP. By telling these stories to students, I start to break down the educator/student dynamic and talk to them as people – equals. This develops relationships with students and is helpful when discussing what may, or may not, work for them. One thing educators must do is open these conversations up so students can unpack their perspectives and the perspectives of others; I believe more harm is done by not talking about these things. So, if you are reading this and thinking "this isn't appropriate for the classroom" I would reply "tell me why you think so". As educators of PMP I believe we must start having these conversations. In my opinion, with truth comes an authentic perspective of reality. Welcome to the *real world* my friend.

REFERENCES

Anthony, B. (2015). Creative conceptualisation: Nurturing creative practice through the popular music pedagogy of live recording production. *International Association for the Study of Popular Music, 5*(1), 139–156. DOI: 10.5429/2079-3871(2015)v5i1.9en

Anthony, B. (2019). *Perspectives on learning popular music production in higher Education from "both sides of the glass"* (Doctorate of Education Thesis, Griffith University).

Berklee. (2021). Andy Edlestein associate professor (online testimonial). Retrieved from https://college.berklee.edu/people/andy-edelstein

Goold, L. (2018). *Space, time, creativity, and the changing character of the recording studio: Spatiotemporal attitudes towards 'DIY' recording* (Doctorate of Philosophy Thesis. Queensland University of Technology).

Green, L. (2002). *How popular musicians learn: A way ahead for music education.* Surrey: Ashgate.

Kolb, D. (1984). *Experiential learning: Experience as the source of learning and development.* Englewood Cliffs, NJ: Prentice-Hall.

Lebler, D. (2006). The master-less studio: An autonomous education community. *Journal of Learning Design, 1*(3). DOI: 10.5204/jld.v1i3.31

Luttrell, B., Goold, L., & Ward, A., (2020). The MacGyver approach: Teaching economy of availability in tertiary music education during covid -19. *Social Alternatives, 39*(4), 27–32.

Songkick, (2013). *T.I. "Whatever You Like" captured from The Live Room* (YouTube video). Retrieved from https://www.youtube.com/watch?v=NuqVKRT66ls

NOTE

1 Post-production rooms are similar to a small project studio, they include a DAW, midi controller, speakers and (in some cases) a very small amount of outboard gear.

7

Popular music production learning approaches

Historically, there have been two primary ways by which PMP is learnt: by "getting access to some equipment and learning through trial and error; and [by] observing someone who knows what they are doing [and] getting them to explain, and subsequently copying them" (Zagorski-Thomas as cited in Bell, 2018, p. 179). Learning has occurred in industry apprentice placements, post-secondary education like professional schools[1] and higher education programs, through the use of computer-based systems and engaging with semi-professional avenues such as books, magazines, the Internet and YouTube videos that fuel DIY applications. As a result, the education process is often implicit and learning may occur when there is no intention to learn or awareness of learning happening (Eraut, 2000). Subsequently, there is great disparity in both the educational approaches and learning realities within the options listed above. In this chapter I will highlight the literature and learning approaches that I have found to have utility with PMPP in higher education.

DEVELOPING PROFESSION-BASED MODELS FOR HIGHER EDUCATION

The problematic relationship between learning in an industry-based apprenticeship and learning in the classroom is undeniable in a world of project studios and DIY releases. Considering this, Porcello (2004) suggests:

> In essence, two different modes of learning are being mobilized: apprenticeship mobilizes what might be called 'embodied learning', while classroom instruction, involves the directed presentation of information by an 'expert'.
>
> (p. 754)

Porcello promotes embodied learning but also identifies the difficulties of applying this in the classroom. He suggests that learning in these instances needs to replicate the activities of the profession. Subsequently, pedagogical designs need to evolve and engage students in profession-based scenarios that prepare graduates for the actuality of working in the music industry.

DOI: 10.4324/9781003055099-10

In synergy with this, Pras (2016) advocates for profession-based targets such as "preparation, management and artistic direction of recording sessions" (Pras, 2016, p. 28). These are the tacit production skills that are best learnt via immersion in many different professional contexts. The term *professional practice* connotes a diverse array of meaning in reference to PMPP including that of DIY applications. Adding complexity to this, as Bell (2018) states, is that the DIY studio, as a music-making entity, can be conceptualised as functioning in at least two different modes: "the do-it-alone (DIA) studio and the do-it-with-others (DIWO) studio" (this chapter). Contemplating PMP creative practice in this way informs the possible scope of social dynamics that PMPP approaches can target. Bell's (2019) "Song Machine Project" is an example of this where students engaged with a tracker/top-liner production process within a higher educational context. Bell's study found that:

> The primary benefit of this pedagogical approach is that 'music technology' is learned within a music-making context reflective of real-world practices. Students experience firsthand the challenges and complexities of collaborative music production practices in which nuanced formative assessment is intertwined in the process.
>
> (p. 4)

I have also recently authored two publications that outline the utility of the tracker/top-liner process in the classroom. This module involves small groups (two to three students) undertaking autonomous work, online/remote collaboration, peer reflection, mentorship and self-reflection on practice and learning (see Anthony, 2020; Anthony, et al. 2020).

The overarching intent of these kinds of pedagogical approaches is to embed profession-based realities into higher education models. King's (2016) interviews with producers enabled him to develop a profession-based framework when designing tasks for students. Seen in Figure 7.1 is King's framework for recording studio pedagogy:

King's findings elucidate the importance of the human perspective of music production: the social engagements, decision making and aesthetics. They insinuate the importance of communication skills and the vibe through the use of the term "environment". King also suggests a knowledge of skills, tools, and practices that are important to capture performances and promote workflow. King's work is aligned with the detailed discussions I have outlined in the front section of this book regarding PMP skills, creative practices and music production culture elements.

Shulman's (2005) work into "signature pedagogies" resonates with King's approach. Shulman explains that signature pedagogies are "types of teaching that organize the fundamental ways in which future practitioners are educated for their new professions" (p. 52). These approaches bridge professional practice with higher education by targeting concepts that challenge students to think, perform and act with integrity, and by so doing, replicate the beginning of a career. In these situations, students are made accountable (remember this word) to educators and peers and are expected

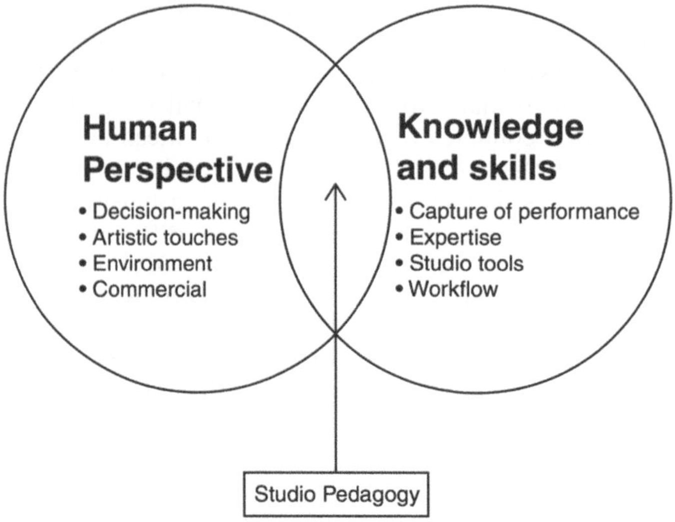

Figure 7.1 King's framework for recording studio pedagogy (Diagram reproduced from King (2016), Studio pedagogy: Perspectives from record producers. In A. King & E. Himonides (Eds.), *Music, technology, and education: Critical perspectives* (p. 63). New York: Routledge).

to participate and make relevant contributions. This kind of work aligns with historical contexts of learning PMP via immersion in a professional setting and also beyond that, within freelance production work, where the learning becomes part of the production process.

Music technology

Music technology often places a more scientific focus on the use of audio technology[2] and then contextualises this technology in musical subject matter. The subsequent interrelation of creativity and technology is then targeted in educational frameworks. This is discussed by Burnard (2007), who highlights the necessary pedagogical change required to address the specific challenges in this area "…the 'deeper' object of musical learning arises inseparably from creativity and technology as interrelated tools" (p. 38). Boehm (2007) discusses interdisciplinary music education including technology and music and bridging the spaces between these two disciplines. These ideas align with the creative ideals of PMP because these are concepts that are not bound by existing schools of thought and are more comfortable within the approaches of students who, themselves, are creative practitioners (as opposed to technologists). In this landscape, these individuals are educated to understand that music is personal and subjective and that technology is a tool one can use to explore their predilections. In synergy with these concepts I have discussed (see Anthony, 2018) a working higher education model that educated students in the application of performance with technology whilst mixing.

Folkestad (2006) targets the relationship between technology, creativity and the student in his comparison of formal and informal learning situations and ways of learning. Within these studies, Folkestad highlights the fact that technology is deeply embedded within the lexicon of young musicians' lives, and that there is importance in seeing that creativity has a relationship with technology, or at least understanding that creativity can emerge through technology. Folkestad (2006) also discusses the importance of social and cultural influences on formal and informal popular music education:

> In order to contribute to the attainment of a multiplicity of learning styles and a cultural diversity – in its true and full meaning – in music teaching practice, music education researchers need to be not only doing all kinds of research in the classrooms, but also to be where children and students encounter musical learning in all its various forms outside schools.
>
> (p. 144)

These words resonate with how popular music is written and recorded in the professional world: musicians are often humming or singing or hearing the song in their heads. Producers regularly formulate visions in their minds and contemplate how to design creative practice situations that will result in appropriate musical outputs. The classroom (studio) is not the only place where popular music education needs to take place; outside the classroom should be experienced also, as this would replicate the many music production cultures that exist in the professional world.

Thompson and Mosley (2015) conducted a qualitative investigation into audio education and its purpose and place in formal education. Their work found that students believed audio education was a route to employment. However, both the students and professionals who were interviewed felt students needed more "opportunities to work alongside established expert professionals" (Thompson and Mosley, 2015, p. 5) and this should be offered as part of the program. Davis et al. (2014) investigated the commonalities between Music Technology education and the industry-based internship programs that are fast declining in the music industry. They suggest that:

> [E]ducational frameworks over the past twenty years have often focused on the technical aspects of studio practice at the expense of the social, aesthetic and human skills required by the industry. These formal frameworks often only focus on the transference of knowledge to the individual diminishing or ignoring the important processes of interaction with the participants in the field.
>
> (p. 1)

The works listed above do much to further promote the importance of the interrelation between technology, creativity and the socio-musical practices of PMP.

Popular Music Education (PME)

Current research suggests that the elements necessary to learn PMP can be demarcated into two categories: (1) audio engineering, musicality, programming and critical listening skills that can be learnt and acquired by formal teaching and practice; and (2) the human aspects of PMP such as practice-based concepts that inform a music production culture and can be learnt experientially during the production process (Pras, 2016; Anthony, 2019). These two schemas that may be musically, socially and culturally influenced, outline a quantifiable process by which the creative practice of PMP may be learnt. However, prior research has highlighted the importance of the form that PME takes (Till, 2017) and this leads to investigations into the learning processes of popular musicians.

Research into popular music pedagogy highlights the importance of "prior learning" (Lebler and Carey, 2008), "self-assessment" and "self-directed learning" (Lebler, 2006, 2007, 2008). These studies align with the renowned work of Green (2002, 2006). Green suggests the popular music student experience can be positively affected by many adaptations of informal popular music learning practices. These include: allowing students to choose the music, learning by listening and copying, learning in groups with minimum guidance, learning in unpredictable often haphazard ways and by integrating jamming – listening, playing, singing, improvising and composing (Green, 2002, 2006).

Green (2002, 2006) highlights that it is important that the student experiences musical freedom and is placed within a musical learning environment that is flexible and where the outcome is not necessarily charted from the beginning. These types of approaches link to the phenomenological work of Macedo (2013), who suggests that phenomenology can offer the foundation for the understanding of the teaching paradigm. Again, the goal is to free the student from any preconceived method in the constitution of knowledge. Phenomenological approaches resonate with Ciobanu's (2017) discussions regarding "active-participatory" methods. Ciobanu states that:

> ...active-participatory learning is a personal activity that cannot be replaced by anything, and the learner is considered to be the manager of his/her own learning, of the whole learning process. This becomes a unique and different activity from one person to another, is determined by the personal history of the subject, by the environment in which it originates, and by the social relationships it develops.
>
> (p. 167)

"Co-creation" (Christensen, 2017) can also be useful as a pedagogical tool within the context of PMPP. Co-creation is roughly defined as "the shared production of something" (Iversen & Pedersen, 2017, p. 16) and this something can be diverse in meaning, in sonic output, in creative concepts, and even in learning. Christensen (2017) attests that

> in an educational setting, rather than focusing on the quality of the current end product, it may be preferable to study the process and

> structures enabling development that qualify the student to reach this level of artistic competence in the future.
>
> (p. 219)

Co-creation is useful within class targets and assessments that focus on student *learning journeys*[3] as opposed to marking the quality of a recorded work.

Engaging with some of these concepts as part of a group encourages the development of confidence in fellow students, and "peer learning" is thereby facilitated (Lebler, 2004). Lebler's (2008) discussions on pedagogical approaches such as peer learning and "student as master" (Lebler, 2007) place the impetus on students to take control of their learning and to develop their trade by replicating historical industry processes of learning in workplace apprenticeships. Lave and Wenger's (1991) discussions on "situated learning" include concepts presented in Lebler's work. Situated learning is where "learning involves the whole person; it implies not only a relation to specific activities, but a relation to social communities – it implies becoming a full participant, a member, a kind of person" (p. 53). This kind of approach encourages students to understand the realities of who they are, what they are good or bad at and to recognise what is needed for them to progress.

Kolb's (1984) "experiential learning" is similar. Kolb's model pursues a link between education, work and personal development and this type of pedagogical design aligns with professional approaches in PMP where many parts of the popular musician's life are influenced by musical experience. In the experiential process, "ideas are not fixed and immutable elements of thought but are formed and re-formed through experience" (Kolb, 1984, p. 26). "Action learning" (Biggs & Tang, 2007) is also useful in these contexts because it "involves changing teaching systematically, using whatever on-the-ground evidence that you can so that the changes are in the right direction, [so that] students are learning better than they used to" (p. 43). These types of approaches are useful for PMPP because they necessitate constant adaptation so as to stay in sync with professional practice.

Lebler and Weston (2015) found that it was necessary that "changes in the music industry were responded to by continually adapting both the [popular music] program content and its assessment methods" (p. 135). This form of flexible learning is largely self-directed and Lebler and Weston's review into the teaching framework of a popular music HE program found that students benefitted from industry internships that counted as subjects. It is this connection to professional practice that produces entrepreneurial graduates who are multi-skilled and flexible in their engagement within industry frameworks. Recent studies into PME have also identified that ideologies of "identity, place, function, lifestyle, gender, audiences, amongst others have now become standard approaches to the study of popular music" (Weston, 2017, p. 104). Weston's musicological studies bare direct relevance to the learning of PMP because the producer designs musical outcomes because of a connection to these ideologies and to promote sales. Hooper (2017) also promotes the inclusion of cultural theory to higher education popular music education by suggesting it highlights "the importance of arming graduates with the tools to be sociologically active and aware as commentators, consumers and creators" (pp. 162–163).

Askeroi and Viervol's (2017) "musical listening" pedagogy is facilitated by a three-step process that includes (1) re-production: learning traditional concepts that reproduce existing recordings, (2) reflection: the use of reference tracks for comparison and reflection and (3) re-contextualisation: applying the knowledge of the former two steps to original recordings. I will be discussing these concepts as pedagogical tools more explicitly in later chapters. Bennett (2017) discusses the unique place that creativity, recording, performance and employability occupy within popular music education. After suggesting that most 21st-century popular music uses a computer in some way, he questions "should digital production skills be ring fenced to specialist 'music production' programmes? I suggest not; the 21st century popular musicians have access to digital production tools that 20th century … creators could only dream of" (p. 291). Bennett goes on to suggest that "it is difficult to argue that a hypothetical popular music curriculum that eschews digital music production skills is not hindering its students' creative development" (p. 291). Bennett promotes the development of multidisciplinary graduates who have a broad skill set including DIY practices. To achieve this, production, performance, stagecraft, vocal, songwriting, music industry, and musicological departments should work together within a holistic approach so that the graduates may, in turn, "respond to whatever creative gigs might come their way" (Bennett, 2017, p. 295).

Lebler and Hodges (2017) is a collaborative study that was written by a popular music educator (Lebler) and a Bachelor of Popular Music graduate (Hodges). The graduate suggested that the PMP course she engaged with greatly shaped the way she currently makes music. In particular, the student experience unpacked that the pedagogical design "develop[ed] in its graduates the attributes of self-direction, self-monitoring, autonomy, intrinsic motivation, diverse skills and the capacity to continue learning" (Lebler & Hodges, p. 280). There has also been some high-quality work done by Tough (2013), who unpacks ways popular music pedagogy can introduce students to *Hit Song Science* (HSS), the study of industry trends and how these can influence PMP creative practice.

Lefford and Johansson's (2017) study investigates student perceptions of the producer's multi-faceted role and whether students, in understanding this role, used this knowledge in their productions. The case study found that there was a "gap between how [students] explain their understanding of producing and their abilities to realize projects" (p. 12) and that this might be because students are rejecting the methods offered to them. Lefford and Johansson explain by saying that:

> though students agree that song lyrics are important, throughout all the stages of the study, all participants state adamantly that approaching a vocalist about the lyrics or performance of those lyrics is quite tricky. Only technical issues, like pitch, seem to be comfortably within the producer's prevue [*sic*].
>
> (p. 12)

It is evident that students may require education concerning the constraints of a producer's role and how this is decided in various PMP typologies. Pedagogy must engage students with these difficult conflicts and discussions that professionals regularly experience. These are the soft skills that take time to develop and may be tacitly learnt.

SUMMARY

Well, firstly, congratulations on getting through the pedagogical literature section of the book! I have found all of the above useful whilst designing courses and the literature needs to be tabled to scaffold what is to come in the following chapters. I advocate that PMP educators consider the overarching message of this book that is founded from my research, professional experience and the literature above. That message is: students require foundation learning in the skills and creative practices of PMP so they can take control of, and be accountable for, their tacit learning of the greater practice of PMP (and all that it entails) into the future. In the chapters that follow, and from within descriptors of pedagogical application, I will refer, both directly and indirectly, to the literature above.

REFERENCES

Anthony, B. (2018). Mixing as a performance: Educating tertiary students in the art of playing audio equipment whilst mixing popular music. *Journal of Music Technology and Education, 11*(1), 103–122. doi: 10.1386/jmte.11.1.103_1

Anthony, B. (2019). *Perspectives on learning popular music production in higher education from "both sides of the glass"* (Doctorate of Education Thesis, Griffith University).

Anthony, B. (2020). The tracker producer and top-liner collaborative music production project. In A. P. Bell (Ed.), *The music technology cookbook; ready-made recipes for the classroom*. Oxford: Oxford University Press.

Anthony, B., Auvinen, T., & Thompson, P. (2020). Learning to be a 'Tracker': A pedagogical case study of learning collaborative music production. *Journal of Popular Music and Education, 4*(2), 211–235.

Askeroi, E., & Viervol, A. (2017). Musical listening: Teaching studio production in an academic institution. In G. D. Smith, Z. Moir, M. Brennan, S. Rambarran, & P. Kirkman (Eds.), *The Routledge research companion to popular music education* (pp. 231–242). New York: Routledge.

Bell, A. P. (2018). *Dawn of the DAW: The studio as musical instrument*. New York: Oxford University Press.

Bell, A. P. (2019). Of trackers and top-liners: Learning producing and producing Learning. In Z. Moir, B. Powell, & G. D. Smith (Eds.), *The Bloomsbury handbook of popular music education: Perspectives and practices*. New York: Bloomsbury.

Bennett, J. (2017). Towards a framework for creativity in popular music degrees. In G. D. Smith, Z. Moir, M. Brennan, S. Rambarran, & P. Kirkman (Eds.), *The Routledge research companion to popular music education* (pp. 285–297). New York: Routledge.

Biggs, J. B., & Tang, C. S. (2007). *Teaching for quality learning at university: What the student does* (Vol. 3). Maidenhead: McGraw-Hill/Society for Research into Higher Education & Open University Press.

Boehm, C. (2007). The discipline that never was: Current developments in music technology in higher education in Britain. *Journal of Music, Technology & Education, 1*(1), 7–21. DOI: 10.1386/jmte.1.1.7_1

Burnard, P. (2007). Reframing creativity and technology: Promoting pedagogic change in music education. *Journal of Music, Technology & Education, 1*(1), 37–55. DOI: 10.1386/jmte.1.1.37/1

Christensen, T. N. (2017). Teaching co-creation: Paradoxes in rock and pop ensembles. In T. Chemi & L. Krogh (Eds.), *Co-creation in higher education: students and educators preparing creatively and collaboratively to the challenge of the future* (pp. 205–23). Rotterdam: Sense.

Ciobanu, N. (2017). The applicability of active-participatory methods. *International Journal of Humanities Social Sciences and Education, 4*(11). DOI: 10.20431/2349–0381.0411019

Davis, R., Parker, S., & Thompson, P. (2014). Preparing the music technology toolbox: Addressing the education-industry dilemma. *Journal of Music, Technology & Education, 7*(3), 313–323. DOI: 10.1386/jmte.7.3.313_1

Eraut, M. (2000). Non-formal learning and tacit knowledge in professional work. *British Journal of Educational Psychology, 70*(1), 113–136. DOI: 10.1348/000709900158001

Folkestad, G. (2006). Formal and informal learning situations or practices vs formal and informal ways of learning. *British Journal of Music Education, 23*(2), 135–145.

Green, L. (2002). *How popular musicians learn: A way ahead for music education.* Surrey: Ashgate.

Green, L. (2006). Popular music education in and for itself, and for 'other'music: Current research in the classroom. *International Journal of Music Education, 24*(2), 101–118. DOI: 10.1177/0255761406065471

Hooper, E. (2017). Do the stars know they shine? An argument for including cultural theory in popular music programmes. In G. D. Smith, Z. Moir, M. Brennan, S. Rambarran, & P. Kirkman (Eds.), *The Routledge research companion to popular music education* (pp. 153–165). New York: Routledge.

Iversen, A.-M., & Pedersen, A. S. (2017). Co-creating knowledge: Students and teachers together in a field of emergence. In T. Chemi & L. Krogh (Eds.), *Co-creation in higher education: Students and educators preparing creatively and collaboratively to the challenge of the future* (pp. 15–30). Rotterdam: Sense.

King, A. (2016). Studio pedagogy: Perspectives from record producers. In A. King & E. Himonides (Eds.), *Music, technology, and education: Critical perspectives* (pp. 45–66). New York: Routledge.

Kolb, D. (1984). *Experiential learning: Experience as the source of learning and development.* Englewood Cliffs, NJ: Prentice-Hall.

Lave, J., & Wenger, E. (1991). *Situated learning: Legitimate peripheral participation.* Cambridge, MA: Cambridge University Press.

Lebler, D. (2004). "Get smarter" music: Making knowledge from know-how. *Proceedings of the Effective Teaching and Learning Conference 2004,*

Queensland University of Technology, Brisbane. Retrieved from: http://hdl.handle.net/10072/2913

Lebler, D. (2006). The master-less studio: An autonomous education community. *Journal of Learning Design, 1*(3). DOI: 10.5204/jld.v1i3.31

Lebler, D. (2007). Student-as-master? Reflections on a learning innovation in popular music pedagogy. *International Journal of Music Education, 25*(3), 205–221. DOI: 10.1177/0255761407083575

Lebler, D. (2008). Popular music pedagogy: Peer learning in practice. *Music Education Research, 10*(2), 193–213. DOI: 10.1080/14613800802079056

Lebler, D., & Carey, G. (2008). *Prior learning of conservatoire students: A western classical perspective*. Proceedings of the 17th International Seminar of the Commission for the Education of the Professional Musician (CEPROM), International Society for Music Education (ISME) 10. Retrieved from http://hdl.handle.net/10072/24742

Lebler, D., & Hodges, N. (2017). Popular music pedagogy: Dual perspectives on DIY musicianship. In G. D. Smith, Z. Moir, M. Brennan, S. Rambaran, & P. Kirkman (Eds.), *The Routledge research companion to popular music education* (pp. 272–284). New York: Routledge.

Lebler, D., & Weston, D. (2015). Staying in sync: Keeping popular music pedagogy relevant to an evolving music industry. *International Association for the Study of Popular Music, 5*(1), 124–138.

Lefford, M. N., & Johansson, S. (2017). *The student's view of the producer's role: Analytical methods, interpreting content and practical project management.* Paper presented at the 12th Art of Record Production Conference "Mono: Stereo: Multi", Stockholm, Sweden. Retrieved from http://urn.kb.se/resolve?urn=urn:nbn:se:du-27701

Macedo, F. (2013). Teaching creative music technology in higher education: A phenomenological approach. *Journal of Music, Technology and Education, 6*(2–3), 207–219. DOI: 10.1386/jmte.6.2.207_1

Porcello, T. (2004). Speaking of sound: Language and the professionalization of sound-recording engineers. *Social Studies of Science, 34*(5), 733–758. DOI: 10.1177/0306312704047328

Pras, A. (2016). What has been left unsaid about studio practices: How producers and engineers prepare, manage and direct recording sessions. In A. King & E. Himonides (Eds.), *Music, technology, and education: Critical perspectives* (pp. 27–44). New York: Taylor and Francis.

Shulman, L. S. (2005). Signature pedagogies in the professions. *Daedalus, 134*(3), 52–59. DOI: 10.1162/0011526054622015

Thompson, P., & Mosley, B. J. (2015). *Audio education: Its purpose and place in formal education and the creative industries*. Paper presented at the 26th Audio Engineering Society Conference, United Kingdom. Retrieved from: http://www.aes.org/e-lib/browse.cfm?elib=17866

Till, R. (2017). Popular music education: A step into the light. In G. D. Smith, Z. Moir, M. Brennan, S. Rambarran, & P. Kirkman (Eds.), *The Routledge research companion to popular music education* (pp. 14–29). New York: Routledge.

Tough, D. (2013). Teaching modern production and songwriting techniques: What makes a hit song? *MEIEA Journal, 13*(1), 97–124. DOI: 10.25101/13.6

Weston, D. (2017). The place of practice in tertiary popular music studies: An epistemology. *Journal of Popular Music Education* (1), 101–116. DOI: 10.1386/jpme.1.1.101_1.

NOTES

1 In Australia professional schools (that are government approved) including the School of Audio Engineering (SAE) and the JMC Academy facilitate music production education.

2 Music technology courses often educate students on computer software and audio technology design and construction.

3 A learning journey is a pedagogical activity where importance is placed on engagement with the experience as compared to the quality of a produced musical work.

8

The *places* and *spaces* of popular music production pedagogy

Earlier in this book I discussed the concepts of *place* and *space* regarding PMP creative practice. PMPP in HE currently occurs in university-type settings. Therefore, the places and spaces of these environments are often different than those found in the professional world and this can influence the student experience in both positive and negative ways.

HIGHER EDUCATION BUILDINGS, STUDIOS AND ROOMS

Many students have suggested in interviews and surveys that HE buildings often carry a stigma, associating them with teaching environments as opposed to music-making environments. As a result, students feel a disconnect with their learning spaces. There can be many factors contributing to this: the décor, furnishings and architecture, for example.

> The real world is hard to imitate in a classroom. The closest we got was when we did a recording session at an external studio, just being away from the classroom and school building felt like it was more real.
>
> (International student survey response, May, 2021)

This disparity can become entrenched in the students' minds during their educational experience.

> Like, mental headspace is so important when creating projects … like the sterile environment of the place, like there is not a lot of windows.… We have no disconnect so it is hard to stay focused when we … have our classes in the same place that we come back in and be creative.
>
> (Hayley, EdD interview 1, February 26–27, 2018)

If the *places* that educators utilise for education are tailored to look and feel creative this helps to remove instructional stigmas and promotes the manifestation of classes that are engaging and enriching experiences. This should then flow over and have a positive influence on student autonomous sessions.

DOI: 10.4324/9781003055099-11

Popular music culture versus learning culture

One of my students who engaged in an industry placement with a professional producer suggested: "Just the culture was different, the professional environment was much more engaging.... The fact that you are actually working for an artist that is paying you. The mentality was different" (Thom, EdD interview 1, February 26–27, 2018). This student's statement links back to a connection to the traineeship model and Porcello's (2004) discussions concerning embodied learning as compared to classroom instruction. To target embodied learning in HE experiences, the socio-musical atmosphere or vibe that is synonymous with PMP must be present for profession-based student mindsets to manifest. Students often described a positive experience in programs at my university:

> I feel the BPM [the program] culture is definitely different than any other university, any other ... course culture and I really like it. There are definitely some positives about the environment ... we aren't just doing this for the grades.
>
> (Hayley, EdD interview 1, February 26–27, 2018)

Another student contextualised the program culture into positive learning approaches that promoted creative control:

> [One thing that] would stop the vibe ... [is] not being able to have as much freedom as we do. We get complete creative control which is what we need as [popular] musicians.... You know, we just want to go in there, do our thing. We learn as we go because that is the nature of what we do.
>
> (Lennon, EdD interview 1, February 26–27, 2018)

Lennon also suggested that the program culture promoted a collaborative environment where mistakes could be made and learnt from: "I think it has always been a big part of it for me to be able to fuck up, rectify what I have messed up and then go and tell ... a lecturer ... 'hey, I fixed this'" (EdD interview 1, February 26–27, 2018). These comments reinforce that students value both profession-based and learning cultures that target embodied learning, and that there is utility in stimulating a popular music culture within the students' program of study. The last comment was also one of several in this interview where Lennon felt comfortable in expressing her opinion by using swear words (as do many of the students). I won't refer to these anymore, but I trust it will support my discussions on the PMP lexicon throughout this book.

By facilitating a *popular music learning culture* as opposed to an *instructional teaching culture*, institutions can encourage student growth. Learning advanced concepts in this way is tacit and encapsulated within the many socio-cultural interactions that occur when popular music is made. However, there is an elephant in the room: you can prepare the environment all you want, but if a student arrives with an unsuitable headspace, learning becomes difficult.

STUDENT HEADSPACE VERSUS PROFESSIONAL HEADSPACE: ACCOUNTABILITY

One of the most humorous situations during the research of this book occurred when I questioned international educators about the headspace, accountability and professionalism of their students. Man, did that bring out some stories! I think, to put it into perspective, we should all think back to our own personalities and traits when we were between the ages of 17 and 25, because this is the predominant demographic in PMP programs. Most students are still young when they arrive at HE, and they have been conditioned by their school experience.

> Most people going into undergraduate degrees … are waiting to be spoon-fed…. It still feels like you've got this kind of school based thing of "this is what happens in school" and you've also got this, "I'm paying for this therefore they should be making me educated [attitude]"…. All the education theory about scaffolding and stuff to support them in that, they're not ready for that right at the beginning, but they're also not being prepared for that. They're not being told that that's the process, they're being told in the marketing (and you can't avoid the fact that universities are guilty of this in a lot of ways), "you're going to be taught by the best people, you're going to have the best facilities", it doesn't say, "you're going to have to work really hard to learn stuff. This is all down to you … you've got to pay your dues". But that's where it needs to go.
>
> (Simon Zagorski-Thomas interview, April-May, 2021)

We must also remember our students are popular musicians. At my institution, students have 24-hour access to the studios and often do all-night sessions. They go out to gigs late at night and as a result sometimes they come to class mentally tired or even hung over. Educators must value the students' engagement with popular music culture and lifestyle whilst promoting their diligence as learners. It is a delicate balance.

A common theme that has come up in my research is that of student *accountability*. The theme arose whilst I probed the students about why they hadn't rehearsed in some classes (see Video 1 on the website at www.routledge.com/9780367517632) and how that reflects on their professional practices.

HAYLEY: I know a lot of people fall into the, "Oh, well just do it for folio" [assessment] and I hate that because then you just end up with songs that you are going to need to revisit anyway.
RESEARCHER: So, can I use a word here?
HAYLEY: Sure.
RESEARCHER: Accountability?
HAYLEY: Yeah, Yeah!

(EdD interview 1, February 26–27, 2018)

My research suggests that professional practice bestows an element of accountability on the popular musician, and this should flow through to pedagogical approaches. Accountability was also recognised in the literature (Shulman's [2005] signature pedagogies), and the subsequent association to professional practice came out constantly in students' reflections.

LENNON: I think what's worked is the fact that we have been made so accountable ... if it's shit it's shit.... So, the fact that you've made us so accountable is so beneficial 'cause that's how it's going to be in the real world.... That first week was rough when everyone was just like, "oh, yeah, we'll wing it" and [we] didn't rehearse, didn't organise parts.

RESEARCHER: And what did I do in the tutorial that made you ... guys feel more accountable?

LENNON: You didn't fix it.... You literally let us make the mistakes. It turned out rubbish, it was shit, nothing sounded good.

RESEARCHER: So, I made you face up to your decisions ... which makes you accountable?

LENNON: Absolutely ...

RESEARCHER: But then when you had your live recording assessment [a few weeks later]?

LENNON: Nailed it. But I think that that's a big thing because if you had've made that first one assessment as well, I still think it would have been a similar outcome ... because we hadn't really had that lesson kind of drummed into us [they needed to fail in Week 1 and learn to be accountable].

(Lennon, EdD interview 2, April 24, 2018)

Student accountability will be unpacked at length in its own chapter, however concerning matters of student headspace, I target this with a mantra: "how you practice is how you play". This mantra is something I have appropriated from sports coaching and promotes applying oneself to the best of your ability regardless of context. It is a mantra that I noticed in many of the greats with whom I have worked: no matter what they were doing, they always did it to the best of their ability. They were accountable in everything they did. This was explained to the students when they were confronted with the fact they had not rehearsed and that their attitudes were more akin to that of a hobbyist than a professional (see Video 3 on the website at www.routledge.com/9780367517632).

As is exemplified in Video 3, I am often quite confrontational with my students, but by the time this occurs (2nd and 3rd year) I have developed a rapport with them. I try to do it in a respectful way, but sometimes just giving them a good old *kick up the pants* helps them understand that they need to start to be accountable for their learning and their future. I find a lack of student accountability raises its head in many ways: lack of preparation; being overconfident, egotistical; not a team player or being passive in class and not talking or engaging. These traits are problematic in PMPP because:

> by definition, a production class has to be an interaction. I don't think that there's any possibility it could succeed if you just sat there and listened to somebody talk to you about what you should do or how you should think.
>
> (Andy Edelstein interview, April-May, 2021)

Alan Williams and I discussed dealing with student egos:

ALAN: We have some students who walk in with egos that don't fit through the door!!!
BRENDAN: They don't last very long with me Alan!
ALAN: So, then we would be doing those folks a disservice if we didn't say, "okay, you need to step back and take a realistic picture of what your abilities are and where you are at this stage".

(Alan Williams interview, April-May, 2021)

Ego is an important student trait to deal with as it can be damaging in production sessions. This must be identified as different to professional confidence and proactiveness, both of which need to be nurtured. Another troubling trait I find in students is passivity:

> Students are too quiet, no communication, I had to talk a lot to promote their engagement ... It is as if they were too scared to critique each other's performances.
>
> (EdD Journal entry, Tutorial 1, February 28, 2018)

My students often hold back because they don't feel comfortable critiquing other student's work. This is mainly the result of fear caused by a concern for offending others. Many international colleagues I spoke with discussed similar experiences:

> I try to draw students out, but it's not always easy. This semester there are two students who just won't jump in the dialog, every class it's like pulling teeth. With one in particular, I've started to see that he has good stuff to contribute. This made me more intrigued and a little pissed off because I was like "come on you're holding back exactly what we want to hear!"
>
> (Andy Edelstein interview, April-May, 2021)

Students' passiveness within PMP has a lot to do with what Harris (2010) calls "the confidence gap". He describes this gap as being the link between fear and freedom – freedom in this instance is having the personal confidence in one's ability during music-making activities and how to communicate those ideas. Therefore, the social context within which students are located whilst producing music is an integral, but often difficult landscape, for young people to traverse.

> Urgh, the social component of people getting together and making something together and having to deal with each other, their personalities,

> their artistic ideas, their technical abilities and limitations ... I mean all of that actually feeds into the thing that we hear on the recording in the end and [dealing with all that leads to creativity and innovation] … if art doesn't evolve, it dies.… I think that for a lot of our students, it's so comfortable sitting in one's bedroom saying: "well, I just play this guitar part over and over again until I like it". I think there's such an emphasis on comfort, my students use that word a lot. They forget what's so valuable about performance and spontaneity.
>
> (Nyssim Lefford interview, April-May, 2021)

The way educators can make students engage more proactively (outside of their comfort zone) is to promote the overarching atmosphere of a popular music culture. If class feels like a collaborative musical engagement (like most sessions), students will become more comfortable and can be nurtured to engage more. However, as educators we need to be intuitive to the unique idiosyncrasies of students and the cultural differences of our cohorts.

SAMANTHA: What I do find are cultural differences.… The main one is simply about ambition. I mean, in London … every single one of [my students] would cite an extremely successful individual, whether that was a producer, or a rapper, or a song writer … and say, "I wanna be like that" and they would work and work and work. Just the sheer dedication and ambition was off the charts and yet here [in Australia], that is seen as, you know, showing off or "ooh, don't be too good" … and so what that's [manifested is] a kind of a real sense of pride in not being ambitious, successful, like that's not something to strive for.

BRENDAN: You think it's linked to the whole tall poppy syndrome[1]?

SAMANTHA: It is absolutely linked to the tall poppy syndrome and [education and industry people] ask me "why don't Australian musicians do well overseas…?" Tall poppy syndrome – it's the only reason, it's because they don't aim for it.

(Samantha Bennett, interview April-May, 2021)

"Tall Poppy Syndrome" describes a cultural phenomenon associated primarily with Australia and New Zealand (it is worth noting that similar phenomena can be found in other Commonwealth countries, Scandinavian countries, and some Asian cultures, amongst others). Varying cultures will all provide their challenges. Another challenge is the influence of the social culture of the university itself and how this may infiltrate class environments and affect students' mindsets.

THE INFLUENCE OF THE UNIVERSITY SOCIAL CULTURE

> Like any sort of institution where young people … have this notion of trying to be cool, bullying is a problem … you have to live with these people for a long period of time. You do not want people gossiping about you in the halls … which is bullying [and] is a problem in many situations.…
>
> (Georgio, EdD interview 1, February 26–27, 2018)

The social landscape of a university can provide aspects of bullying and harassment or unethical social situations that may flow over into the class room and have a negative influence on student engagement. These insidious elements must be monitored and dealt with. However, some other social elements, such as strained dynamics between individuals, have also proved useful as learning contexts for students in my tutorials:

> The students did really well today to confront a "people problem", where the singer wanted a different engineer because of personality clashes. They all worked the problem, communicated and got it done.
>
> (EdD Journal entry, Tutorial 10, May 24, 2018)

This particular social conflict from one of my tutorials can be viewed in Video 4, and Elliot had the difficult task of telling Thom he was to be replaced. These tough decisions, themselves important soft skills, must be made for the betterment of the music, and students should be empowered to deal with them during student-led process (see Video 4 on the website at www.routledge.com/9780367517632).

Difficult social dynamics can be turned into learning opportunities: "I think I'm still at the point where I'm learning not to get caught up in other people's shit... it's got to be about the end product, not about how I feel about a situation" (Lennon, EdD interview 2, April 24, 2018). One of the students in the tutorials for my EdD study (JP) was a highly active and employed session drummer, and the students saw his tutorial engagement as a role model. JP was always prepared; during class he exuded professionalism and his actions personified the "how you practise is how you play" mantra. JP had a great influence on other students' headspace. As Georgio stated: "I see him as real-world, like a real-world session would be five of him or four of him or whatever" (EdD interview 2, April 24, 2018). It was beneficial having that kind of individual as a collaborator in the tutorial experience. If they are not present in the cohort, this could be offset by engaging professionals within the pedagogical experience.

Jo Lord had a different but equally interesting perspective regarding the greater university social culture:

> There is also the tricky notion of HE institutions now becoming a service industry, whereby the student is seen as the paying customer. Subsequently, this has given students far more influence power over how content is structured and delivered. It has also led to cohorts containing numerous … entitled students that only seek to engage in what they deem as relevant to them … [being] dismissive of anything they deem as unnecessary, including not engaging in core learning materials and assessments. Yet, they would continue to complain that they are not receiving the 'experience' they are paying for. The nature of the 'business' is impacting pedagogy drastically in the UK and driving many good practitioners and academics out of the field and creating an over-saturated graduate market with qualifications lacking true academic integrity.
>
> (Jo Lord, questionnaire response, April-May, 2021)

"WHO ME?" THE EDUCATOR'S HEADSPACE

I have met and conversed with many PMP educators over the two-and-a-half decades that I have been facilitating PMPP, and I believe I could hand pick the ones who students feel they learnt the most from. There are different ways to facilitate an education; we may have different learning outcomes or different goals in mind for our graduates but I don't really mind those differences since they are all institution or program-specific. The one thing I have trouble hiding my frustration about is an educator with a toxic headspace. Most of these spawn from ego or *fear of being found out*, and they come in many different shapes and sizes: the *Dr. Evil* (their ego wants to make lots of Mini Me's), *The Know-It-All* (is usually an elitist and uses too much technobabble or obfuscatory language), *The Baloney Sandwich* (covers up a lack of experience and practical ability with street talk) and *The Bestie* (becomes too friendly with students). You may think I am being a bit harsh here, but, in my opinion, there has been some poor education offered in PMP in various capacities and I am not alone in that thinking:

> The real world of music production is both shrinking and distrustful of universities. Because we are in a period of transition between people who are trained through the apprenticeship model and people who are coming in with degrees. There are a lot of people coming in with degrees who can't walk into a studio and do stuff, because we've had quite ordinary courses. People doing bad education basically.
>
> (Simon Zagorski-Thomas interview, April-May, 2021)

Alexis Kallio reflected on the educator's role

> I think we are still fighting against the positioning of the lecturer as the expert, [which is evident] in our accreditation system and how educators get rewarded. All these certifications and things, it's like giving people gold stars. It's all designed to quantify an educator's abilities and I think what makes a good tertiary educator is something well beyond that. Like setting up structures for students to engage and lead their own learning which is particularly hard with traditional assessment structures and because of where students were before they come to us. We need to be able to change their [the students] mindsets – and also change our own. Education is an art.
>
> (Alexis Kallio interview, April-May, 2021)

Education should be focussed on the students' needs as popular musicians, not the educator's knowledge or experience. However, the PMP educator is different to many other fields because they are associated with popular music culture. The educator needs to embody that culture to develop trust, respect and a professional relationship with students that balances the educator/student dynamic away from master/apprentice models. Instead, this relationship should resemble peers. This takes time but it is integral for enriching and enlightening learning to occur. Jarritt Ahmed Sheel elaborates:

JARRITT: When I was teaching in the Bronx the last time I taught high school, I started with, 'yeah, you've gotta call me Mr Sheel and blah, blah, blah' Their [school] was very controlled.... And those kids were like, "yo, you're not listening to us" and so I became "Sheel" or "Mr"... you know we had a very on board relationship even when I left the program, it was like "love". It wasn't hate and so because I did that, I think I got more done with them than I could've with all that other power trip stuff, which is really about management.

BRENDAN: Like what you're talking about is that educator-student relationship, it needs to be...

JARRITT: Equal!.... Why can't it be equal? like if you give kids equal respect, what will that do to you? Nothing! No, you've just gotta be nice to them, that's it. Dude come on! And how hard is that? (Jarritt Ahmed Sheel interview, April-May, 2021)

The difficulties in this approach often arise with students because they need to (1) understand this is ok and (2) not take things too far – *The Bestie.* However, music making can be a soul-searching and expressive activity when working with lyrics, musicality, performance, communication, psychology and emotion. By taking the time to develop a suitable rapport with students, the educator prepares the necessary environment to delve into these difficult pedagogical terrains when needed. To do this well, the educator needs to be perceived as being an authentic person.

BRENDAN: Do you think students relate to honesty, the authentic [educator]?

BRYDEN: Yeah I would definitely say so ... I'd argue that students identify falsity pretty easily and, in contrast, value the way in which we are open and honest about their work and that we take time to set up a dynamic in class that makes everyone feel able to share their thoughts, ideas and creative work. I find there is great value in creating authentic environments where we are a community of musicians rather than conforming to the typical lecturer/student dynamic.

ZACK: I do explicitly sit them down when I first meet them and say: "look this is supposed to be fun. We are in a privileged position that we can spend our day doing this. But I am going to talk to you like I would talk to friends and colleagues, so I will swear, I will have a laugh with you, I will slag off something that you've done. You can all do the same to me, it's totally fine. And that's not a "hey I'm Uncle Zac come be my pal sort of thing", it's not that, it's going back to that authenticity thing that you say, it's like this is who I am, I get employed to stand in front of you for a reason, I'm not going to pretend.

(Bryden Stillie and Zack Moir interview, April-May, 2021).

The development of meaningful working/learning relationships in popular music cultures requires individual authenticity and this is so important.

> We are real humans, we're not going to pretend that we're not and let's all get on together.... We bake that in, we get to know them [students] really well, they come and chat to us.... We might be sitting in the canteen at lunch time and they'll come over and ask us questions and chat to us. It's real human education, it's not us and them. I think that's so important.
>
> (Zack Moir interview, April-May, 2021)

The ideals Zack, Jarritt and Bryden have discussed represent important advice for PMP educators. The creation of popular music is inclusive of the culture that surrounds it, and this culture relies heavily on the development of authentic relationships between people. Education in this realm must reflect that culture. Relationships need to be formed and equality needs to be established. By founding PMPP in these relationships, student abilities, both inside and outside of the classroom, can flourish.

TECHNOLOGICAL SPACES

The students of PMPP need to be developed into flexible and versatile practitioners so they can go on to portfolio careers or become innovative music makers. The type of technology that institutions use is not necessarily a dealbreaker for me (virtual, hardware, etc.). Rather, the student's flexibility as a practitioner is key. I am a firm believer in placing emphasis on the actions of the user within the *technospheres* (see Chapter 3) producers operate as opposed to the specific device that is used. Professional producers work and adjust their actions instinctively according to what they are hearing whilst using any form of technology. Their actions are reactive to the sound of equipment, and students should learn to do the same. Andy Edelstein agreed and elaborated:

> The gear doesn't really matter. It's what's in your head. There are extraordinarily successful engineers and producers who don't really understand how their tools work but they have a vision and can twist knobs until they realize that vision.... The tools were largely an impediment to getting their vision realised. So, I think we can look at all these technological inventions as being blessings and curses.
>
> (Andy Edelstein Questionnaire reply, May, 2021)

Workflow is also important, and tactility must be a consideration for all music producers. I hate using a mouse when working; I like the tactility of faders. However, when discussing this with my students, some of them suggested they find the mouse a very creative and tactile device. Educators must value the workflows that students find creative and help them to unpack why this is the case. This is no more prevalent than within the debate regarding DAW software:

> For a very long time it has been argued that Pro Tools is the 'Industry standard' and thus should be the system of choice for the program. In my experience, this argument has gotten weaker every year. We have students who increasingly use Logic or Ableton, which they have used since high school for their work …
>
> (Nyssim Lefford interview, April-May, 2021)

I have a moral issue with any software company that does not make their platform readily accessible to all students of PMP, therefore I do not base my pedagogy on software that is device specific (Mac or PC). Many of our students come from low socioeconomic backgrounds and they may not have the funds to purchase high-end computer technology or expensive software licences. Therefore, the PMP courses that I currently convene recognise Pro Tools and Ableton Live as core to the work we do, but we have versions of Logic Pro and other software on campus and allow students to operate on any software with which they feel comfortable for their assessment work. Subsequently, I promote the software Reaper to many of my students as a highly affordable and useful home-studio alternative. Although it is not financially viable to supply students with every possible piece of software they may want to use, variety is important because different types of technology are strong in varying creative practices. For example, Pro Tools is based on a mixing console, so it is very useful in pedagogy that focusses on recording and mixing applications. However, as a compositional tool, I find Pro Tools' MIDI and programming functionality is not as user-friendly as Ableton Live. Of course, many educators will need to become proficient at more than one software application, and in doing so, educators can be seen to exemplify the skillset of modern professionals. Institutions must assure that the technological affordances (*technospheres*) and the creative spaces that they provide are most suitable for students to undertake all elements of their course work as well as allowing them the freedom to innovate, create and pave the way for future music making.

SUMMARY

I have outlined above many considerations of place and space in PMPP contexts. I do think it is easy to make excuses and blame HE frameworks for the limitations we face in this regard (especially headspace) because it takes an enormous amount of energy to action change in these areas: to provide the most suitable facilities and nurture a popular music culture and vibe in the spaces within which students work. Educators must target these areas and challenge their institutions to allow them the freedom to do whatever is needed in order to promote a high-quality student experience. Without a suitable overarching context that considers the necessary culture, places and spaces for PMP to occur, it will never live up to the promises of real-world experience.

REFERENCES

Harris, R. (2010). *The confidence gap: From fear to freedom.* Melbourne: Penguin.

Porcello, T. (2004). Speaking of sound: Language and the professionalization of sound-recording engineers. *Social Studies of Science, 34*(5), 733–758. doi:10.1177/0306312704047328

Shulman, L. S. (2005). Signature pedagogies in the professions. *Daedalus, 134*(3), 52–59. doi:10.1162/0011526054622015

NOTE

1 The tall poppy syndrome is the cultural phenomenon of jealous people holding back or directly attacking those who are perceived to be better than the norm. It is associated with both Australia and New Zealand.

9

The cohort (both sides of the glass)

One of the overarching recommendations of this book is that PMPP's more advanced concepts are facilitated via a *both sides of the glass* perspective, and that students develop an understanding of PMP practice by designing, implementing and working in various music production cultures. This recommendation, therefore, emphasises the interaction of all people involved in music-making activities (artists, writers, programmers, engineers) whilst learning occurs, and the diversity (acoustic, electronic and online collaboration) that is present in popular music. This activity may be nuanced to suit the various programs around the world and the musical styles of the students involved.

POPULAR MUSICIANS AND TECHNOLOGISTS WORKING AND LEARNING TOGETHER: HOLISTIC PEDAGOGY

I have a long association of facilitating PMPP for a Bachelor of Popular Music Program that involved an integrated cohort of producers, artists, engineers and performers. Following a recent campus relocation, I currently convene a series of music production courses at my institution. Following my request, upper management at our university approved a series of first- and second-year foundational creative studio practice courses which include both music technology and popular music major students working side by side. I am compelled to advocate the value in this integrated approach (in both foundational and advanced frameworks) because it is an efficient, affordable and effective way to facilitate education. The advantages of this method are threefold: (1) all students develop a holistic foundational domain of PMP knowledge that will support their future endeavours as portfolio musicians and producers, (2) knowledge transference encapsulates a broader scope (musical and technological) due to the tacit learning that occurs through the engagement of students with diverse skillsets and (3) the social skills of students (the soft skills that take longer to develop) are targeted in a passive way from the beginning of the higher education experience and given time to develop. A further outcome

DOI: 10.4324/9781003055099-12

is that students often view themselves in new ways following this kind of experience.

ZACK: I think that the more we specialise, the more we're creating a situation that's artificial … I know we've seen students come in thinking that they're one thing and just grown when they've encountered … other ways to think about music, so the cross section is just incredible, and I think the more we can do to foster it the better.

BRYDEN: The intersection of all these different experiences [performer, producer] is where the students start to see some realities and when they can actually start working … thus breaking the barriers because they don't really exist anyway.… I think the way that our course is set up is to try to understand the opportunities rather than define a path that they are definitely going to go down … there are so many creative possibilities.… I think the job of the establishment is to give you the opportunity to try it out, to test if that's something that helps you blossom as a musician.

(Bryden Stillie and Zack Moir interview, April-May, 2021)

By integrating cohorts into a single learning experience, pedagogy bridges to modern applications and diverse learning ensues through immersion in the collaborative process. "As a pop student, I loved the combining of CMT [music technology] and pop students as we all learnt from each other in really collaborative ways" (Student experience survey response CSP 1, 2021). However, this method is not without its challenges. Some educators I discussed this with suggested that varying cohorts require different learning outcomes.

> The technical rigor is different from the performance rigor which is different from the record production rigor. The result usually ends up with some portion of the triumvirate disappointed because their focus is not at the forefront of the conversation. As much as I prefer having all three parts present, explaining the compression timing of Optical vs. VCA is usually a glaze over moment for the uninitiated; even though I think I'm communicating in a way that anyone can relate.
>
> (Prince Charles Alexander interview, April-May, 2021)

Prince Charles Alexander correctly presents one of the main obstacles I encounter with this approach, namely, that students are often not as enthusiastic regarding applications with which they may not be specifically interested. In these situations, it is the responsibility of the educator to use a lexicon that will connect with all students, and to mentor students regarding the importance of foundational domain knowledge that generates multidisciplinary graduates and promotes their professional independence and longevity. Andy Edelstein continued this debate:

> Not sure populating a traditional course with both would work since the needs of each cohort would be considerably different, though it's an

> interesting concept since it presumably could provide each an opportunity to better understand the perspective of the other.
>
> (Andy Edelstein interview, April-May, 2021)

The point Andy makes at the end is key to this discussion: "an opportunity to better understand the perspective of the other". These collaborative engagements fuel tacit learning of diverse content, they facilitate social interactions between students from differing backgrounds which are the beginnings of networking and the development of communication skills. With the ever-eroding divide between producer, engineer and musician and the ubiquitous nature of the portfolio musician in today's industry, it is time educators engage with approaches that consider PMP as a holistic learning framework. To do otherwise places students into outmoded past practices where roles were separated. Sam Bennett expands on this concept:

> I don't see that you can really delineate. We're in 2021! All their music's gonna end up on Spotify or YouTube, so there has to be some form of technological mediation. You could even break it down even further and ague that instruments are music technology and that performance itself is technologically mediated, the vocal chords are a technology. Therefore, I just think that to say over here we've got our musicians, over here we've got our composers and here's our performers and here's our recordists [is incorrect]. It's not separate groups and I don't think it can be.
>
> (Samantha Bennett interview, April-May, 2021)

It is time for pedagogical approaches to be brought into alignment with the current state of play in the industry where PMP roles are blurred and interchangeable, and diversity is commonplace.

PEDAGOGY THAT IS EQUITABLE FOR TECHNOLOGICALLY AND MUSICALLY MINDED STUDENTS: STUDENTS DESIGNING THEIR LEARNING

An integrated approach does not mean that all students must target the same achievements in advanced concepts (mixing, engineering, performance, etc.). The development of foundational technological and socio-musical knowledge in early years allows the students a larger and more professionally focussed skillset to develop. The emphasis shifts to student-centred learning approaches and students becoming accountable for their professional development whilst tailoring their advanced learning outcomes to their interests. Performers may discover they want to engineer and mix their own product. Music technologists may discover a passion for songwriting and performance; educators, meanwhile, can monitor individual student growth within mentorships and advise of specific autonomous targets should the need require.

Within integrated cohorts I have found that students not only learn by doing but they learn by watching others – songwriters and musicians learn about sound and technology by watching and communicating with sound engineers and sound engineers learn about lyrical messaging and performance techniques by working with skilled performers. The students who took part in my doctoral research confirmed this. Lennon stated: "It was really cool sitting back and watching it all happen 'cause you can take heaps in that way" (EdD Focus Group 1, May 29, 2018). By placing students like Hayley (who was not strong in sound engineering) in a music production culture, she learnt about sound engineering via immersion in the process, being around it while it happened promoted: "just little moments where you're just thinking outside the box, like sound engineering creatively was also really cool" (Hayley, EdD Focus Group 1, May 29, 2018).

Another key aspect in this approach is the tacit learning of communication and people skills that happens in the socio-musical interactions during classes. Also, educators must equally value the musician's development of technological knowledge with the sound engineer's development of musical and performative knowledge. Too much emphasis (in my opinion) has been placed on education that focusses on technological capabilities as opposed to socio-musical workflows. I questioned many educators as to their opinions on having musicians and technologists working together whilst learning how to generate a performance and develop their interpersonal skills:

TOBY: Well, it's tough.... I mean what we do a lot is we just bring in pros from around town...

BRENDAN: Cool. And how's that taken with the students?

TOBY: They love it because they have spent so little time seeing someone on the other side of the glass perform at these levels. But I will say that ... having students perform that aren't necessarily great is also an opportunity 'cause what they will find is when the pros come in, well that was easy. You just point a mic at that thing makin' sound and it sounds great. But [if] you get ... more amateur musicians ... you have to work for it.

(Toby Seay interview, April-May, 2021)

Lachlan Goold agrees:

> I started recording punk bands that had only one rehearsal a week and had little knowledge of music theory. Some of these bands required a great deal of work to make the music sound good and it is from this place that I learnt many performance improvement techniques.
>
> (Lachlan Goold interview, April-May, 2021)

Within integrated cohorts, students can experience what it is like to work with musicians from all performative levels and thereby develop techniques that they can later use. The musicians bring genre-influenced affordances that engineers can experience whilst working; many of my students have learnt how to produce and record vocal sessions for Pop, Hip Hop, Rap, Folk and alternate Rock and Punk vocalists. I would postulate that this experience has not only engaged my students with diverse and explicit engineering skill applications, but, and perhaps more importantly, given them experience in working with people from diverse cultural backgrounds. At the same time having performance students work with a variety of engineers allows them to contemplate the psychological and technical skills that aid in generating high-quality performances. An approach that facilitates integrated cohorts and targets foundational domain knowledge followed by student-centred learning in later years is highly advantageous when contemplating a graduate's integration into a portfolio career. Bryden Stillie elaborates:

> We make sure students … get the opportunity to develop skills that they need and want to develop and because of that they get a really good idea of where they want to go as a musician. Rather than us kind of saying, you're a producer, you're a composer, you're the drummer. It's like well in reality going back to the kind of employability side of things, everybody has got a bit of everything now days. So, we provide foundations in early years but not all students will continue [in the same way] … which is pretty cool.
>
> (Bryden Stillie interview, April-May, 2021)

PRIOR LEARNING

My research has shown that PMPP students require significant prior learning to engage effectively with the more advanced aspects of PMP in practice-based activities (Anthony, 2019). Prior learning, in this case, helps to locate the pedagogy on a particular foundation and is specific to all PMP skills, as outlined in Chapter 1. This enables students the ability to move into the actuality of *doing* PMP in diverse music production cultures. For example, in my doctoral study, Georgio stated: "sound engineering, we would've been like screwed had we not known what we learnt in previous years about that" (Georgio, Focus Group 1, May 29, 2018). It is only through obtaining these sound engineering skills that students could stop focussing on them and integrate more with the music, their surroundings and their production ideas. Students could also rely on others for the skills in which they were lacking. Having songwriters in the classes allowed the tacit awareness of melody and lyrics to inform the group's aesthetic production decisions. Integrated collaborative environments also informed students of the interconnectedness of the PMP skillset. Georgio commented on how he learnt that high-quality sound engineering affected a performance: "it made her go, 'ahh', [by adapting the sound] and she was like playing better, singing better" (Georgio, Focus Group 1, May 29, 2018). However, educators need to consider the innovative abilities of their students and value the different types of skills they may bring to the classroom. As Alan Williams states:

> I also think that we have to open up curriculum to be a little more welcoming of what the students already walk in knowing. Because I think there is sometimes, a workflow that students develop that might

> be foreign to their faculty who, you know, still have scars from editing ripped tape!
>
> (Alan Williams interview, April-May, 2021)

Modern musicianship

Regarding matters of musicianship and music theory, I do often contemplate whether some of those in control of the institutions and the education that we facilitate are (1) governed by a very specific outlook on what education should look like, (2) naïve, and simply not aware of the foundational musical knowledge that is synonymous with popular music or (3) of the mind that popular music as a form of practice is worth less than the Western classical canon. For 20 years I have experienced the insipid equality wars between rivalling HE departments of music (classical, jazz, pop and music technology, for example) that are often the result of elitist behaviours from both staff and students. Social equality in all musical styles is imperative for music education to move forward in a meaningful way. No musical application or style is more important than any other, they are simply different. Popular musicians may not be most suitable in classical contexts and classical musicians may struggle in popular music contexts.

> I'll never forget … trying to make a rock album at university and getting some cellists in to do some recording. I brought them in … and I said to them "hey can you riff over this guys?" They looked at each other, saying "what?" I gave them the chords and said "off you go". And you just sit there saying, "hang on a second, what are they teaching them up at the other building? This is crazy!"
>
> (Russ Hepworth-Sawyer interview, April-May, 2021)

Yet there seems a definitive lack of institutions targeting popular musicianship and musicality with their students:

> The way the university does it [foundational knowledge in musicianship] is the first year is a core year where everybody takes the same courses. So, they're doing general integrated musicianship. So, it pisses me off because [the students of PMP] they're doing like Bach cantatas on the piano when I want them to be doin' like ragtime, you know! [PMP].
>
> (Jay Hodgson interview, April-May, 2021)

Simon Zagorski-Thomas added the influence of the school system to this debate:

SIMON: Formal education at school level is still much more focussed on classical music than it is on pop music, although it's in there. I don't know if you've ever looked at secondary school syllabus about popular music, it is shit.

BRENDAN: Are we in transition though?

SIMON: It feels like a glacially slow process. There is a really interesting argument in America, a guy called Phillip Ewell gave a talk [that said] the way we've developed music theory has been purely about promoting the values of Western classical music. On stupid obvious levels like improvisation verses notation. Improvisation is not seen as a skill that classical musicians should acquire, but notation is seen as a skill that popular musicians should acquire because that's a real skill.... But what it's saying is the idea of diversity in music is not about going to find black classical composers and putting them into the canon, it's about saying music theory needs to theorise all music and that means there are going to be lots of different music theories that should all be equally valuable.

(Simon Zagorksi-Thomas interview, April-May, 2021)

To their credit, many institutions (like mine) are attempting to be progressive in this regard by opening music degrees to reconsider musicianship and theory courses. I spoke with Chris Stover about this:

BRENDAN: So when considering the tacit knowledge of popular musicians: improvising, songwriting, musicality and performance. Are institutions setup to deliver these?

CHRIS: Most institutions as they currently stand are not set up at all to handle that. The conservatorium we work for is making really interesting progress in that direction, and should be a good case study in the very near future; they have a lot of different types of students – classical, jazz, popular music, music theatre, and so on – but they are all in very tightly sealed silos. But it comes down to tacit knowledge and what are expected to be the kinds of knowledges and skills you need to get into the program in the first place, and if reading music and understanding 18th- and 19th-century harmony or sight reading is the benchmark a lot of really good musicians are going to be left out. Similarly, if we flipped that model and made understanding acoustics and manipulating frequency envelopes the benchmarks, another bunch of students would be marginalised. The problem comes when we keep repeating this utter failure to communicate between these different areas.

(Chris Stover interview, April-May, 2021)

Chris went on to explain that most institutions are resistant to interdepartmental communication because of unfounded fears regarding legacy and the loss of rigor. He suggested that music theory class could be a good starting point.

> Maybe it should be the place that all the students can come together and hash out ideas and learn from each other. But then a concern can arise from the more conservative musicians who say "but that's going to water the theory class down, students won't get the rigorous attention to canonic music that they need to study." I am firmly convinced this doesn't need to be the case. I am developing a music theory curriculum

> that resonates with all music students, where all music students start together in a mixed cohort and do a year of broadly-oriented, exploratory theory before branching off into specialised knowledges. So instead of beginning with harmony, tonal function, solfège, start with cross-cultural and cross-genre concepts of pitch, rhythm, sound, timbre, and so on. This way all students are learning outside their specialist areas, and rather than reproducing old models, contributing to the production of new knowledges.
>
> (Chris Stover interview, April-May, 2021)

The practical outcome of Chris' approach would be that, as cohorts progress through the university, their engagements would break these silos down by cutting across genres and mindset. Therefore, the networking of string quartets, jazz musicians, pop singers and songwriters would then continue into the future, and this would create richer networks and help reshape the industry itself (Chris Stover interview, April-May, 2021). The academic and professional industry are in need of reform because popular musicians, for some time now, have been occupying spaces where their sense of self-worth has been questioned.

> Yeah that value system is inherent in the way that we teach all music. A guy called Nicky Brown was the music director for London Community Gospel Choir, he'd been taught entirely in Church, he couldn't tell you the name of any chords, but he was an amazing arranger.... I remember him saying to me, "I wish I'd learned theory", and I said "your theory is so far in advance of mine as far as harmony is concerned".... He had this sense of insecurity and inferiority.... he felt that kind of sting that a proper musician "knows notation".
>
> (Simon Zagorksi-Thomas interview, April-May, 2021)

Popular music instruments

The definition of popular musical instruments also requires constant consideration. Popular music makers are using all kinds of devices, MIDI controllers and laptops as performative mediums. Worldwide there are many aspiring popular musicians who cannot read a note of music but are capable of programming and performing a complex piece of music, and these skills need to be recognised as both valuable and suitable for student admission to popular music programs. Simon Zagorski Thomas commented:

> There isn't a quick way to fix that.... what really needs to happen is the people who have got power in this regard, like the RMA[1] in the UK, the AMS[2] in America, they need to open up the doors to it. I know that John Stewart down in Brighton, he's just started working with the RMA about having a popular music branch. In 2020 this is the first time that the RMA have decided that they ought to have one. It's very much like the way things like artificial intelligence slowly crept into Oxford,

> Cambridge, Edinburgh, all those places. It's because all the old guys slowly died and got replaced by young people! They're not going to give up their power by saying "actually you're right, I'm rubbish". The whole thing about privilege is that … giving up costs you something and very few people want to do it. [Philip Ewell is a good example] he was not talking about popular music as such, he was talking about the way that he exists in the classical world.
>
> (Simon Zagorksi-Thomas interview, April-May, 2021)

STUDENT AGE, MATURITY AND MATTERS OF EQUITY AND EQUALITY

I have had students enter first year aged 16–65. I place minimal restrictions on the capabilities of students due to age; many of my mature age students have integrated into their cohorts, done very well in their studies and gone on to have successful and meaningful futures. I do, however, suggest we need to consider student maturity within the pedagogical work we do. I have alluded previously in this book to important social learning processes that may involve students unpacking their own personality, reflecting on emotional responses and engaging in socio-musical collaborations and conflicts that can be difficult at times. Much of this work is made easier by students who have developed their social maturity, and this often comes as students get older. My doctoral research found there are some associated realities that must be considered with this style of pedagogy. By unpacking personal issues, students can become emotional about their inadequacies and this can be a confronting experience for young people to deal with (Anthony, 2019). This suggests that social maturity may benefit education in the more complex psychological aspects of PMP. Students from third year onwards, in my experience, have been better equipped for this kind of pedagogy to take place as compared to first- and second-year students.

Equity, inclusiveness and diversity in the cohort

In my opinion, PMPP needs to be equitable, inclusive and a viable option for students from diverse origins. It is therefore problematic, as I found through my discussions with Adam Bell, if students can only access PMP courses by studying classical, jazz or non-popular music programs. This is because many students from low socio-economic areas are self-taught and do not play classical music. Unfortunately, bias is also present in the education system; international surveys of students confirm this: "some educators come off as sexist, condescending, and didn't like to facilitate learning. They'd favour the male students in the class" (Student international survey, 2021). Most educators I interviewed also agreed that PMP currently operates in a very white- and male-dominated industry (both academically and professionally), and agreed that this is a complex problem that exists in many fields necessitating social change. Alexis Kallio suggested that "music students, because of their skills in listening (especially), expression and collaborating, can lead social change in a way that is more difficult in other disciplines" (Alexis Kallio interview, April-May, 2021). However, these are complex issues to solve and change must be approached from many perspectives. For example: there is a lack of diversity in our cohorts and educator ranks. In my interview with Kirsten Hermes, we discussed the potential that integrative cohorts had in promoting equity, diversity and inclusiveness in PMPP.

KIRSTEN: I think that us merging performance, production and business cohorts will really help with gender bias because, at the moment, most production students are male. I think in the (our) new first year, all of them are, which is terrible, isn't it?

BRENDAN: Well, why does that happen?

KIRSTEN:... I think there's an ingrained culture where girls just don't feel like doing those things ... and the issue with that is that in order to succeed in the music industry, it's very important to have interdisciplinary skills. If you're a singer and you cannot do any of the technical stuff, it's going to be tough. I've met a few female students that were scared to even try working with technology.

BRENDAN: I've gotta ask you – is that the reason: that they're scared?

KIRSTEN: I think there's a lot of different factors at play here.... I do think that ... there are microaggressions, sometimes the boys are more confident and I think it's generally like a cultural thing where men are sort of taught even though they feel insecure, to sort of go for it anyway ... whereas women kind of were taught, "oh, they know exactly what they're doin' and I don't, so therefore I must just be really bad at it" and I think both men and women need to work together to resolve that. Like, when I was a student, I felt it was very much "read between the lines", no-one would say you can't do this because you're a woman, but you feel like there's less eye contact being made. Like, it can be really very small subtle things that just send this message of you're not really part of the club.... So I think the only way to change that is to really manifest a different situation and I think having a course that has no strands and that forces everyone to do everything ... hopefully can help resolve that.

BRENDAN: Yeah and empowerment in the students that everyone's capable of anything.

KIRSTEN: Yeah. Exactly ... also encouraging different ways of doing things is important ... there is a way in which men culturally tend to approach these things in education where it's kind of about the "gear" and about the "mix settings" and the "technical specifications" and then the women ... tend to think about it in a more audio-visual or kind of creative way.

BRENDAN: So, as educators, can we look at how we deliver our content in a different way to target that imbalance?

KIRSTEN: Yeah, I think we have to.... Actually, what's really interesting is that in my university, the students are with one [male] lecturer for sort of two years and in the third year they have me and that has caused real culture shocks to some people because I do tend to teach things

> in a more conceptual way. So, if they're all boys in the first place and they're taught by an older male character that's been very successful [producer] in the past ... and then they come to the final year and get me ... they go, "are you actually a lecturer?, 'Cause you look like a student!" And they just think "this person doesn't teach me what I'm supposed to know because the previous person has been quite dogmatic and has said, "here's the ten steps to mixing'"... if I then come in and say, "depends" they just think, "oh, I'm not being taught!" So, I think it's about just making it much more mixed right from the start, mixing the students together, mixing the boys and the girls together and mixing us lecturers and just making it clear to them there's various schools of thought – they all can work, they all have their place in the world, but it's not like one is better or more advanced or one teaches you more.

BRENDAN: They're just different.
KIRSTEN: Yeah.

(Kirsten Hermes interview, April-May, 2021)

Kirsten's words identify the positive influence that integrated cohorts (both sides of the glass) could play in matters of equity, diversity and inclusiveness. These types of matters must be targeted immediately; it is alarming that, as an educator, Kirsten has suffered microaggressions from her students. My discussions with Amandine Pras suggested that dogmatic educational approaches that emphasise technological applications as opposed to aesthetic considerations and critical thinking only exacerbate the gender imbalance that we see in PMPP (see Fisher & Margolis, 2002; McNally & Pras 2021).

> Amandine: [In the past I have been] under pressure to deliver trade-school-type education that is cheaper to deliver than university-type programs.... As an example, there is pressure to teach courses that focus on DAW functionalities [compared to] teaching critical thinking and critical listening. I believe that these types of production courses are 1) not giving the best education to students and 2) reinforcing gender inequalities since studies in computer science education showed that teaching technical functionalities without putting them in context played a big role in demotivating non-male students.
>
> (Amandine Pras, questionnaire response, April-May, 2021)

If educators and the institutions for which they work aspire to deliver courses that are equitable for all popular musicians then the ways in which courses are delivered must embody the most suitable options for a modern, inclusive and diverse cohort. Courses like this should recognise the innovative musicianship, performative and technological skills that students bring to the classroom as relevant, and class activities need to subsequently promote their development. Institutions that open their curriculum to support student-centred approaches such as this would be paving the way towards positive social change in the music industry.

REFERENCES

Anthony, B. (2019). *Perspectives on learning popular music production in higher education from "both sides of the glass"* (Doctorate of Education Thesis, Griffith University).

Fisher, A., and Margolis, J. (2002). Unlocking the clubhouse: The Carnegie Mellon experience. *ACM SIGCSE Bulletin, 34*(2), 79–83.

McNally, K., & Pras. (2021). *Audio+: A research forum to transform or traditions*. Proceedings of the 2021 AES International Conference on Audio Education. Retrieved from https://www.researchgate.net/publication/353342642_Audio_A_Research_Forum_to_Transform_our_Traditions/link/60fc1aa80c2bfa282afb7826/download

NOTES

1 Royal Musical Association, see: https://www.rma.ac.uk/

2 American Musicological Society, see: amsmusicology.org

Part 3

Pedagogical approaches

This section presents recommended pedagogical approaches for PMP programs and courses.

DOI: 10.4324/9781003055099-13

10

Popular music production pedagogy

An engagement with diverse music production cultures

My research to date has suggested that effective learning of PMP involves more than enabling students to produce a high-quality piece of music or becoming proficient in the skills of music production. More importantly students developed an understanding of the holistic practice of music production which, when assimilated by students, could be tailored towards their specific aspirations. Students acquired these abilities by working in, and reflecting on, diverse music production cultures (Anthony, 2019).

An import conclusion drawn in my doctoral research was the great value that students attributed to learning in diverse, collaborative, technically challenging and open-ended settings. This was exemplified in some of the participants' word art diagrams which were collected in the doctoral study, seen in Figures 10.1 and 10.2. The diagrams were drawn by the students during two separate focus group sessions and were in response to questions I asked about PMP. In the first focus group, students were asked to rate the skills of PMP in order of importance and the results were intriguing. After about five minutes of working on this, the students drew a diagram on the whiteboard that showed all of the skills as interrelated; one was only as important as the others that influenced it. This diagram was adapted throughout the hour-long discussion and the focus group experience is captured in the picture shown in Figure 10.1.

This activity brought out some wonderful discussions during its application and it seemed the participants were proud that they had worked out a different way to think about the question:

LENNON: Can we have more than one in the same spot?
ELLIOT: We should have like a Venn diagram.
ALL: [Laughter]
GEORGIO: Or like a web of sorts.
HAYLEY: Get creative people in here and this is what we're going to do to it.
ALL: [Laughter]
LENNON: Fuck yeah!
ELLIOT: Outside the box! That's the point. We're going there.

(Focus Group 1, May 29, 2018)

DOI: 10.4324/9781003055099-14

Figure 10.1 represents the students' interpretation of their engagement in the course *Music Production 1*, and resonates in some ways with the systems model of creativity (see Chapter 5). The students have instinctively drawn three elements that form a system: a *learning* element (domain knowledge), a *real-world* element (the field and further domain knowledge) and an *agency* element. Of note is how various applications like live recording are situated in all three elements. In this way, the three elements fuse together in the middle and are often interrelated and informed by one another.

In the second focus group, students were tasked with drawing a picture which illustrates an effective PMP educational experience (see Figure 10.2). Many of the individual elements in this diagram ("Turbo Charged", for example) will be discussed in the following chapters of this book.

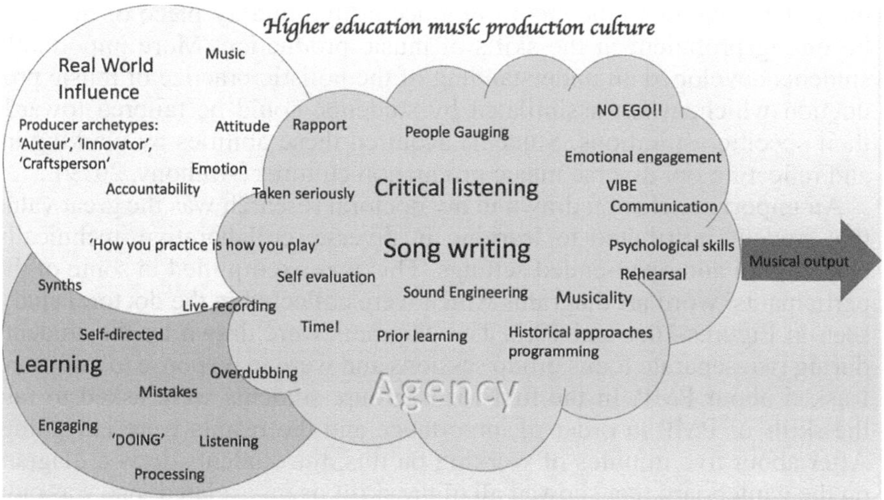

Figure 10.1 Student's word art (recontextualised) – higher education music production culture (Anthony, 2019, p. 149).

Figure 10.2 Student word art – higher education popular music production pedagogy (Anthony, 2019, p. 195).

However, upon reflection, students suggested the holistic practice of PMP is best learnt from within collaborative, profession-based tutorials where all aspects are interrelated and experienced in practice-based scenarios. It seemed that by educating students in this way, they learnt that everything gravitates around the song and the socio-musical engagements of the people making the music. Students realised that the skills and creative practices were all interrelated and were influenced by each other. This represented an understanding of the practice of PMP that could then be applied to a specific type of music and nuanced for its purpose. Bennett (1980) attests a similar perspective about learning to become a rock musician where a group collaborative experience is vital. He writes:

> [T]he situations in which their music is created and performed and the meanings which come to be attached to the innumerable immediate events of their musical endeavors [sic] are, then the fundamental elements of knowledge that rock musicians share with one another.
>
> (p. 6)

Bennett then goes on to suggest that musicians generate a collective definition of situations that they treat as real, and therefore they produce the experience of being real. This also occurred within the diverse music production cultures with which my doctoral students engaged in their tutorials. Through collaborative practice-based engagements, the students developed a collective understanding of the holistic practice of popular music production. This is a macro outcome; the thing we are trying to achieve in general terms. On more micro levels this understanding of practice is personalised by students as they engage with collaborative environments and unpack their learning with the group. The relationship between the macro learning (the practice of PMP) and the micro events that help build it (personalised learning strategies, informed practice and practice-based engagement) are exemplified in Figure 10.3.

HOLISTIC ENVIRONMENTS THAT NURTURE PERSONALISED LEARNING (MICRO 2)

The pedagogical planning of my PMP courses adjusts to any current shifts in the industry (technology, consumer platforms, genres, etc.), but the intricacies of the music production cultures that students engage with are diverse from year to year as a result of individual students' musical preferences. This presents an opportunity for the educator to plan their content, but to remain instinctive and to let the pedagogical framing adapt to the students' musical and stylistic preferences. A holistic pedagogical experience (where strategic goals and outcomes are targeted) that also nurtures personalised learning is the result. The process promotes that students develop a personalised version of the practice of PMP; something unique to them, the music they make and the aspirations they may have. Students also work with the music of their peers and this helps to broaden all students'

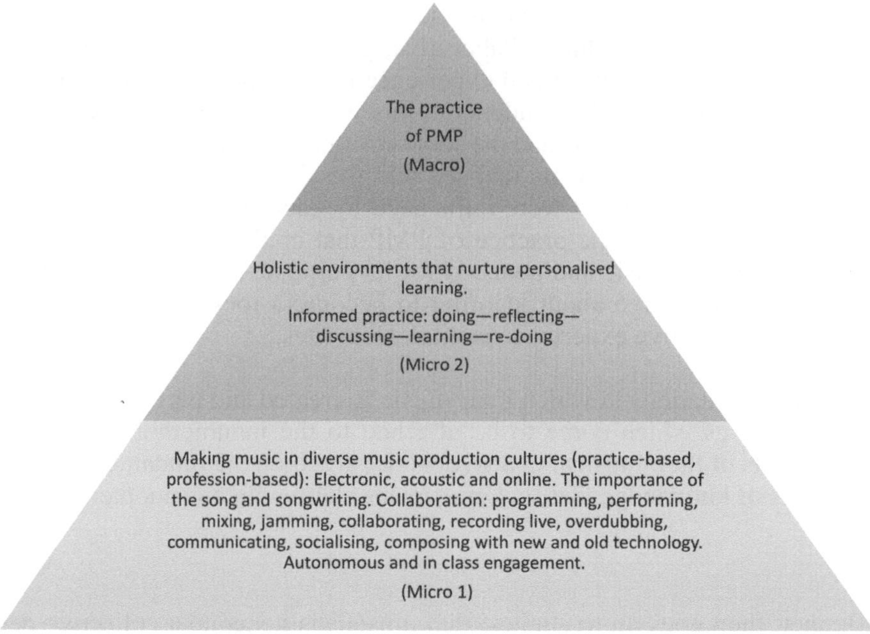

Figure 10.3 Pedagogy that develops a holistic understanding of the greater practice of popular music production.

understanding of the practice of PMP from within diverse contexts. This way of working eschews the possibility that degree programs will only manifest cookie-cutter graduates that primarily tailor to the current trends of the industry. Instead, graduates are empowered to target current trends but also, should they choose, emphasise the development of their own artistic identity. It provides clarity for graduates: they get to work out who they are and what it is they are trying to achieve with music.

INFORMED PRACTICE: GATHERING KNOWLEDGE, DOING, REFLECTING, DISCUSSING, LEARNING AND RE-DOING

If we dial down a little further, personalised learning is facilitated through reflection and discussion with peers. This process often happens after music making occurs, and students of my doctoral research suggested that open discussions and feedback sessions with other students regarding tutorial-based activities allowed for reflection, understanding, re-contextualisation of future possibilities and the subsequent development of students' confidence with practical applications. Simon Zagorski-Thomas clarified the utility of reflection with peers following practice-based processes.

SIMON: I think experiencing and then reflecting on it and reflecting on other people's experience as it's happening [is important] because its different if you're in a class with some mates who are doing something badly [compared to] if you are being shown something by a lecturer. That's a different type of experiential process.

BRENDAN: So … reflecting together, what went good, what went bad and talking that out with peers is helpful?

SIMON: Yeah. Again, led by staff because they're not going to do it automatically, they've got to be kind of led into that process. That's the way to learn, especially with stuff that is practical. Even with theoretical subjects the way to do it is to make them think, not to make them know something.

(Simon Zagorski-Thomas interview, April-May, 2021)

Pedagogy where students are making music and then reflecting, analysing and evaluating together (critical inquiry) is something that I strongly advocate for. It helps to take the onus off of the educator's opinion (which students can come to rely on) and promotes that the students become educators themselves. "I mean we have to embed, you know, responsibility and challenges and maybe even insurmountable challenges [because] the students need to learn about education as much as they need to learn about production" (Rob Toulson interview, April-May, 2021). These types of approaches resonate with Lave and Wenger's (1991) situated learning and Toulson and Hepworth Sawyer's (2018) discussions that contextualise Bloom's cognitive domain into music production learning journeys:

> The journeys taken by a student in order to develop experience and skills in the field of music production align especially well with Bloom's cognitive domain [exemplified in Figure 10.4] (Bloom et al., 1956)…. Bloom et al. described that the reception of knowledge alone only leads to shallow learning, and that deep learning is only developed by moving up through the cognitive domain, through comprehension and the effective application of knowledge. Equally, engaging in reflective analysis is essential for practitioners to learn from the application of their knowledge.
>
> (Toulson &Hepworth-Sawyer, 2018, p. 273)

If students incorporate Kolb's (1984) experiential learning within Bloom's cognitive domain then an experiential method of learning is facilitated:

> Kolb's cycle of Concrete Experience [includes] Observation & Reflection, Formation of the abstract and general concepts, then the synthesis of Testing implications and concepts in new situations. Through this cycle, the learner can reflect and begin to develop and implement their own original ideas and autonomy by ultimately demonstrating true wisdom in their field.
>
> (Toulson and Hepworth-Sawyer, 2018, p. 280)

I have provided a class video (see Video 5 on the website at www.routledge.com/9780367517632) that exemplifies the above discussions in action. In this video, the students engaged in a live recording and used

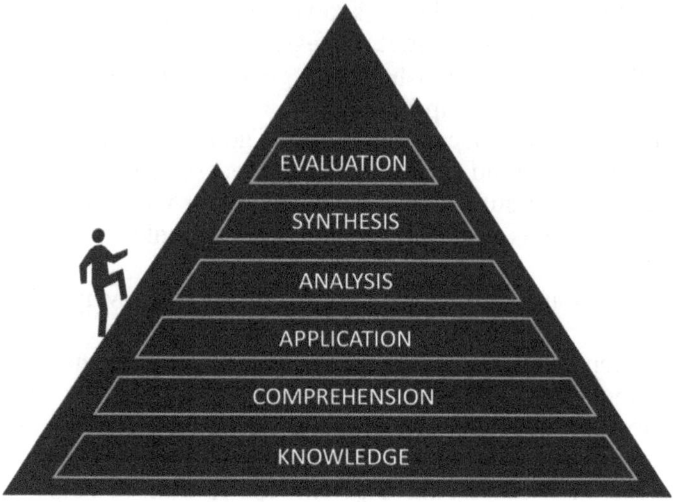

Figure 10.4 The six levels within the cognitive domain of Bloom's taxonomy (Diagram reproduced from Toulson, R., & Hepworth-Sawyer, R., (2018). Connected learning journeys in music production education. *Journal of Music Technology and Education.* 11(3), p. 273).

student observation and reflection within a practice-based tutorial to move through the cognitive domain, as seen in Figure 10.4. You will note that, in this video, I outlined the learning framework at the beginning of the class – doing (playing), listening and reflecting. Throughout the session my footprint was quite strong as a facilitator, but it was required because this was a new environment for these students (their first time in this type of context). Student feedback in this tutorial occurred between recording takes and then in a final feedback session at the end of class. This feedback session unpacked PMP practice and there are definitive moments where students reflected on learning explicitly (see 7: 10). This class was not a part of my doctoral work and, as such, the video was ad-hoc, so I apologise for the poor camera angles in advance. Also, the music students played was a cover so, due to copyright restrictions, I have muted the audio whilst students are playing. I will discuss later why students performed a cover and not an original work. Nevertheless, the importance in this video lies within the students' words as they are working through the process: doing, listening, reflecting, analysing, evaluating and learning together.

PROFESSION-BASED ACTIVITIES: MAKING MUSIC IN DIVERSE MUSIC PRODUCTION CULTURES (MICRO 1)

Profession-based activities (like the video example above) where students engage with music production cultures can be extensive (working on the same song over many weeks or until completed) or, in contrast, can be one-off sessions (beat creation, pre-production, lead vocal or heavy metal guitar tracking or mixing). However, this rigorous and practice-led approach must follow more formative classes (skill-based education, the ambit of first year), and so is usually restricted to final-year students. It is also important that the educator's role evolves from one that exemplifies practice in second year to that of a music-making peer and facilitation in final years. These sessions mostly occur in the classroom (studio), but students can also be tasked to engage with music production cultures and feedback sessions in their autonomous work since many music-making scenarios are difficult to facilitate in large group sessions – electronic music, programming sessions, for example.

The following videos show sessions from my doctorate which students felt bridged education and professional worlds. The first such example can be found in Video 6 which facilitated a tracker/top-liner session (see Anthony et al., 2020). This is an assessment item where students collaborated via online and face-to-face means whilst producing DAW-based songs (Pop, Hip Hop, Rap, etc.). This video shows students working together to diversify a student's top-line for their tracker song. I often integrate this activity into the tutorial experience so I can monitor student progression. It is a long video that also exemplifies the student's use of humour, linking to lyrical messaging and the utility in designing creative social cultures. The specific element of this class that student's felt bridged to the profession was when they were separated to work on a top-line melody (3:00–10:00) (see Video 6 on the website at www.routledge.com/9780367517632). Another example is video 7, a live acoustic recording where performers and engineers worked together to generate a linear high-quality performance (see Video 7 on the website at www.routledge.com/9780367517632).

SELF-LEARNING: WHAT STUDENTS BRING TO THE ENGAGEMENT

Some of the most significant and meaningful education earned through working and learning in music production cultures involves discovery of oneself. Successful music producers are forever nurturing the music-making experience, but to do so they need to be acutely aware of their own persona, vibe or traits. Skilled music producers adjust their personality to suit the needs of the people making the music. Within pedagogical contexts, self-discovery can be quite confronting for students. My doctoral studies contained student engagements that were uncomfortable communicative situations, but these facilitated great learning for students if the opportunity to discuss and unpack these things was provided.

In one of my classes I was discussing with my students how to read people's body language and facial expressions and how to use psychological skills to promote high-quality communication. When students engaged in this process, an uncomfortable situation arose as the result of poor communication. Within this conversation, Thom's critique of Elliot's piano part was strongly worded, and Elliot's rebuttal equally so (see Video 8 on the website at www.routledge.com/9780367517632, 1:18-2:18). This conversation rendered the room silent; the studio space was uncomfortable until I changed the direction of the discussion. Later in the class we discussed how the students can learn from an experience like this (a personality clash) and how they need to consider their own personality traits when communicating with others (see 3:39). It was highlighted to the students that professionals engage with conflict as part of creativity regularly and that they learn to deal with it in different ways. All of the above is exemplified in an in-class moment from Video 8.

PROJECT MANAGEMENT

An understanding of the practice of PMP has little utility unless it is authentically applied. I support pedagogy where students get the opportunity to engage in PMP from within the context of project management. In this setting, PMP practice is undertaken alongside communication with clients; musical brief consideration; budgeting; scheduling; booking of talent, engineers and session musicians and meeting deadlines. These additional applications are an integral element of the greater practice of PMP.

> For me everything is music, but you have to develop people skills and understand music from different cultures and you have to make it [translate]…. [So] I took a standpoint [to other teachers] I don't think we should train our students for a traditional vocation in the media industry … I think we should study production, media production and music production as a field of practice, and the most important skills the student should learn are [located within] project management…
>
> (Sören Johansson interview, April-May, 2021)

Mark Marrington explained how industry partners can be used to increase the profession-based context of such an experience:

> During the year we have students doing projects. There's a big project module they do across all three years where they work with different people in different parts of creative production practice and [the industry partners] set the client briefs to respond to and they come in and feedback [is given on projects]. [This includes] people from broadcasting, people from the dance music industry etc. you name it….
>
> (Mark Marrington interview, April-May, 2021)

Project management engagements (where students are working with clients from the industry) have utility as profession-based pedagogical approaches. My student surveys included opinions that suggest that local industry is valued by students:

> Please have engagement with the local music scene. A University out of touch with the local scene will result in an empty studio. History of Western Art Music is fascinating but irrelevant to the history of Digital Audio Arts.
>
> (International student survey response, May, 2021)

> I had one audio professor in school who was also a professional engineer for major country artists, and he gave the best challenges, the best advice, and hooked me up with the internship that started my career. I think the curriculum could use much heavier focus on creativity, emotion, performance, psychology, and actually being a working professional.
>
> (International student survey response, May, 2021)

These types of opportunities would seem to be the stepping stone for students' integration into the current industry. They may also provide opportunities for students to consider what is possible but not yet available. This could help students think, dream and manifest future industry frameworks.

INDUSTRY INNOVATION

I want to conclude this chapter by discussing the issue of empowering students to target industry innovation. In my opinion, the current structure of the popular music industry is not supportive of all working musicians: corporate leadership has let many musicians down, and governments have devalued the importance of music and the arts more generally in our society for some time. Therefore, I often feel compelled, as a part of learning the practice of PMP, to prepare graduates to be the next innovators of the industry. Chris Stover and I discussed the dichotomy of working within a system in order to change the system:

CHRIS: I would also like to be moving into a space, as a society, where musicians do not have to be constantly scrabbling for little bits of money here and there … to meet their basic needs.

BRENDAN: So, in a way, if we are considering "being entrepreneurial" as an attribute, we are developing students for the current social structure of the world we live in, and perhaps that is not the optimum structure. You know what I am saying?

CHRIS: I like that very much.

BRENDAN: … It's like "The Hunger Games". We currently live in a situation that is not optimum, so we feel duty bound to prepare students for that which they will experience but at the same time many of us believe change is required [and want to action that change].

CHRIS: Exactly. So, they have to be happening concurrently. You have to work within the system but the system doesn't care! [laughs]. So all you are doing is hurting yourself if you don't work within the system. I say this carefully, because I also believe there's a real truth to the notion that you can't use the master's tools to dismantle the master's house, but in a really important sense, you have to work within the system if you are going to have anything to do with changing it.

(Chris Stover interview, April-May, 2021)

As educators we must help students prepare for industry integration. However, we must also value their transition from learners to industry innovators in a hope that they will make suitable changes required in the industry. Educators must also nurture those who will not become working musicians and use music as something they love to engage with in an ancillary fashion to other career choices. In these diverse educational landscapes, the educator's duty then becomes one of student development and empowerment.

CHRIS: As much as they might think they do, students don't really know much when they are eighteen years old. I know I didn't! But it's not like we know more than they do, we just know more about how little we know, if that makes sense. And we've been "not knowing" for a lot longer! And [students need] experiences and failures and redirections and all of these kind of things, like just helping them to think about things that they would not have ordinarily thought about ... we are trying to broaden their ability to imagine what can be done in the future.

(Chris Stover interview, April-May, 2021)

If students develop a rigorous understanding of PMP as a form of agency (including project management), they can then be challenged to broaden their perceptions of how that practice can be used into the future to achieve the innovation they feel is important. As such, students would be given all the necessary tools and experiences to develop into innovative popular music practitioners.

REFERENCES

Anthony, B. (2019). *Perspectives on learning popular music production in higher education from "both sides of the glass"* (Doctorate of Education Thesis, Griffith University).

Anthony, B., Auvinen, T., & Thompson, P. (2020). Learning to be a 'Tracker': A pedagogical case study of learning collaborative music production. *Journal of Popular Music and Education, 4*(2), 211–235.

Bennett, H. S. (1980). *On becoming a rock musician*. Amherst: University of Massachusetts Press.

Bloom, B., Englehart, M. Furst, E., Hill, W., & Krathwohl, D. (1956). *Taxonomy of educational objectives: The classification of educational goals. Handbook I: Cognitive domain*. New York: Longmans.

Kolb, D. (1984). *Experiential learning: Experience as the source of learning and development*. Englewood Cliffs, NJ: Prentice-Hall.

Lave, J., & Wenger, E. (1991). *Situated learning: Legitimate peripheral participation*. Cambridge, MA: Cambridge University Press.

Toulson, R., & Hepworth-Sawyer, R., (2018). Connected learning journeys in music production education. *Journal of Music Technology and Education, 11*(3), 269–286.

11

Learning frameworks

It is important that lectures, tutorials and student autonomous engagement work together to achieve overarching pedagogical goals. Therefore, I target the development of the student's knowledge of the greater practice of PMP through the concept of "sensemaking" (Quintana & Fishman, 2006). Sensemaking requires that the understanding of terminology and reasoning necessary to engage in a practice be developed before engaging in the basic practices of an activity. I have found in my classes that if students are also given an opportunity to reflect together, sensemaking can be achieved within the group and on individual levels. Every educator can consider overarching approaches such as sensemaking and the ways in which the different educational experiences available to them can be utilised to that end.

LEARNING OUTCOMES

Learning outcomes are integral to the design of educational experiences and assessment; they lay the platform for the various elements of sensemaking to occur. The Association Europeénne des Conservatoires (AEC) describes learning outcomes as "the statements of what a student is expected to know, understand and be able to do at the end of a period of learning" (2017, p. 3). Given the diversity of music programs that currently facilitate PMPP and the diverse social, cultural and musical affordances of their student cohorts, learning outcomes need to be explicitly considered for every course. However, there are some categorisations that can prove useful in aligning international approaches with learning outcomes. These are: practical (skills-based) outcomes, theoretical (knowledge-based) outcomes and generic outcomes (ibid., p. 4). From within these, the AEC suggests that learning is described as acquiring:

- skills: described as cognitive (involving the use of logical, intuitive and creative thinking) and practical (involving manual dexterity and the use of methods, materials, tools and instruments);
- knowledge: described as theoretical and/or factual;
- competences: described in terms of responsibility and autonomy.

(ibid., p. 4)

DOI: 10.4324/9781003055099-15

Examples of these three types of learning outcomes (spanning undergraduate to doctoral classifications) are profuse on the AEC (2017) report and I promote these as influence in the development of courses. Some examples of learning outcomes for Bachelor level study include:

Skills: Communicate information, ideas, problems and solutions to both specialist and non-specialist audiences through a range of media and presentation formats.

Knowledge: Identify a range of strategies to interpret, communicate and present ideas, problems and arguments in modes suited to a range of audiences.

Generic: Respond creatively and appropriately to ideas and impetus from others, exhibiting tenacity and the ability to digest and respond to verbal and/or written feedback.

(p. 6)

Even though consistency in the use of learning outcomes would do much to align international approaches, there is no one-size-fits-all approach. Moreover, individual institutional autonomy and diversity is important. The AEC insist that institutions "rightly claim the ability to determine what is important to them in light of their unique national and institutional perspective, mission, vision and context" (2017, p. 10). As such, the findings of the AEC could be used to inform the development of learning outcomes that reflect institutional values but address course requirements from multi-faceted contexts (skills, knowledge and competencies). If learning outcomes are utilised in this way, the pedagogical experience can challenge students to operate as both practitioners and critical thinkers.

ON DEMAND, BLENDED AND FLIPPED LEARNING

I am writing this book from home, not only because I am fortunate enough that my employer is allowing me to do so, but also because my country is currently in COVID-related lockdowns. I don't necessarily want to date this text, but the one thing COVID has helped us do is open our minds to possibilities in the virtual world. However, online education existed long before we all went into lockdown. Aspiring professionals have been engaging in autonomous learning via online resources such as *Pensado's Place* (2013). In this YouTube series, Dave Pensado uses an interview process in which watchers can stream programs live and ask industry professionals questions related to their methods. Many musicians can learn PMP skills by watching Internet videos; it seems pertinent however, to question the validity of some of these frameworks because, in contrast to HE's use of online delivery, the posting of YouTube videos is not a scrutinised educational framework and can be designed and marketed by individuals who have little or no industry experience or educational background. However, there is value in utilising this kind of resource:

> What's interesting about the shift in delivery system … over the past ten years with YouTube … [is students] can get the technical knowledge

> through watching all the right YouTube channels.... Actually that kind of training session, where you're in a group with people and you say, press this button and do this, we don't really need to do that anymore. As long as we organise education we can use the technology that's there to bypass all those things that are really demonstrator's jobs. The trouble is that universities are a bit scared of that, because they don't know what that means, they also think that students feel hard done by if they're just telling me to watch YouTube.
>
> (Simon Zagorski-Thomas interview, April-May, 2021)

Streaming classes, online courses

Modern students are used to watching video-styled education and educators should adapt to this and utilise blended learning to their advantage when necessary. I have no problem directing my students to creditable online resources, but I also believe there is an opportunity for HE institutions to develop innovative online content. During a snap lockdown in 2020, some of the foundational tutorials on sound recording and instrument knowledge I convene needed to be facilitated online. The sessional staff involved took action and, instead of recording videos, they streamed classes from their home studios live on the Internet with high-quality audio. These classes were interactive and allowed students to type questions into YouTube during the live stream after which discussion sessions were held on another video platform (zoom, for example), allowing further questions to be explored and answered. An example of these classes can be seen on Videos 9 and 10 (see Video 9 and 10 on the website at www.routledge.com/9780367517632). Of note in all video resources is that effort must be applied to making the videos as engaging as possible; the educators have used multiple cameras and are using a lexicon that is representative of popular musicians. Other classes have been pre-recorded and supplied as video resources for students to watch prior to face-to-face engagement (see Video 11 on the website at www.routledge.com/9780367517632).

These kinds of approaches are not new; many institutions have developed online courses. Andy Edelstein reflected on the utility of these.

> I wrote our online courses Pro Tools 101 and Pro Tools 110 which are sanctioned by Avid for their certification program... those are fully self-contained [and] I've seen the outcome of those courses, it's more than enough for skill acquisition.... This semester [during COVID] they allowed us to offer campus students access to sections of online courses while we're still remote. I can easily imagine integrating such resources into our campus curriculum. Though expensive to develop over the long run, it's worthwhile.... My third Berklee online course is called Microphone Techniques ... I don't think it will ever need to be updated. Is someone going to come up with another drum mic'ing technique? I don't think so! I'm just glad at least one of my online courses doesn't need a major overhaul every couple of years.
>
> (Andy Edelstein interview, April-May, 2021)

Herein lies the problem with virtual resources: any video/pre-recorded content may require updating which potentially results in an ongoing expense. Regardless, COVID resulted in the proliferation of virtual classrooms and many, including myself, utilised video resources as on-demand learning frameworks (Voss, 2018).

I currently utilise blended and flipped learning for first year foundational activities but prefer face-to face for the advanced, more socio-musical contexts in later years (excluding online music making). Flipped classrooms require students to watch video resources before coming to face-to-face lectures and tutorials to engage in content. These video resources are made available to students so they can watch at their own pace and on numerous occasions should they feel the need. The videos have proven far more efficient and effective than standing in front of a class and talking or exemplifying a practice in a face-to-face tutorial. Students are at different levels; they forget things and efficient teaching is not embodied by an educator repeating content in numerous iterations of tutorials when content can be accessed and watched as an on-demand resource. However, there are limitations with using on-demand video resources: "In these remote versions of online campus courses I'd say that we're missing something in not being face to face if the type of production that you care about is collaborative production" (Andy Edelstein interview, April-May, 2021). Video resources, and even to some extent synchronous online classes, lack the face-to-face interaction of students, the intuitive ability to change tact should students' questions require and online classrooms manifesting a different type of social context as compared to being in the same room, collaborating and making music together.

I have been fortunate that my institution has invested the money to help me develop a blended learning resource for a first-year foundation course in music recording principles – DAW and acoustic recording foundational knowledge. The focus of this book is not on foundational courses, however, the added utility of this approach in developing students' prior learning justifies some description here. The course (CSP-1) currently includes a weekly pedagogical video that students watch prior to the week's engagement, a 1.5-hour lecture/listening/feedback session (all students) and a 1.5-hour practice-based tutorial (8 students per class in various studio locations). I try to promote student engagement in my video resources by utilising acting and humour and ensuring a suitable length of approximately 15–20 minutes.

An example of one of the CSP-1 videos that focusses on mixing is available at Video 12 (see Video 12 on the website at www.routledge.com/9780367517632). The 2021 cohort consisted of both popular music and music technology major students, and the course has been successful. Student experience survey results (2021) achieved a mean of 4.8 out of 5 and students' comments supported the pedagogical approach that incorporated the use of the on-demand video resource.

> The flipped learning was really helpful, as I was able to work through the video content at my own pace, and then get more out of the lecture and practice-base tutorial because I already had some understanding.
>
> (Student CSP1- SEC response 2021)

> I loved having the videos to go back to. It's helpful if you've missed something along the way.
>
> (Student CSP1- SEC response 2021)

Bryden Stillie also uses flipped learning in PMPP approaches and vouches for the utility of these:

> A significant number of our music production and studio-based modules are taught using a flipped classroom approach. This enables students to learn concepts before coming to class. This means that students are primed with the knowledge of the concepts and the theory that surrounds them.
>
> (Bryden Stillie interview, April-May, 2021)

Blended learning in the foundational applications of PMP skills develops the initial elements of sensemaking: understanding the terminology and different types of applications for a practice.

THE PURPOSE OF LECTURES: KNOWLEDGE, LISTENING AND FEEDBACK SESSIONS

Lectures should be short enriching experiences. I have learnt the hard way that making my students sit through two hours of me dictating to them about PMP has limited utility. Learning about popular music comes from interaction and I love having discussions with students about the music they are getting into at the time, listening to it and then chatting some more. Students can learn so much in these settings, sometimes even when they are not aware learning is occurring. Paul Thompson said something intriguing about student learning:

> I ask my students, where does learning happen and they say it happens in the lecture room and the control room and I go man, I'm going to blow your mind, it happens in your brain. When you learn something new, you create new neural pathways and the like. That means, this changes your brain composition forever and that's huge!
>
> (Paul Thompson interview, April-May, 2021)

We must be creative in how we communicate with our students (like Paul's comments above). Subsequently, popular music educators should use the term *lecture* loosely – if they do not, students may run for the hills! However, by using this term, students will understand it connotes a practice where people congregate and content is listened to and discussed. I have supplied some exemplar lecture topics in Appendix B that include some of the more advanced concepts of PMP like vision and the producer archetypes. However, lectures must be interactive and engaging, if possible, there should be jokes, lots of videos and music played and lots of back-and-forth discussion between the educator and the students.

Lectures can increase student's domain knowledge by playing music from the Billboard charts, modern, vintage and alternate sources. Techniques (modern and vintage) used by musicians and producers should be identified and analysed and this should inspire conversations regarding integration of these applications into student work. These discussions may lead further, for example, general discussions regarding the popular music industry itself. Students should be immersed in inclusive discussions where all subjective opinions are valued but where the educator can infuse and contextualise facts (ribbon microphones were used in 1940s music, for example). The lecture experience is then removed from images one might connect to studying algebra or the periodic table of elements. Lecture content can definitively be planned, set out in PowerPoint presentations and target the specific content and learning outcomes of the course. The delivery of the lecture, however, needs to be engaging, inclusive of student interaction, nurturing of the breaking down of the educator/student dynamic and, if possible, be fun.

Listening and feedback sessions

I am one of the first to complain about poor listening environments in lecture rooms but listening goes further than sonics (yes really). For too long the term "critical listening" has been normalised to revolve around frequency response, compression and spatiality. That's the last thing I want my advanced students focussing on! I want them listening and talking about the lyrics; the song; the vocal delivery; the space created by musicality and instrumentation; the energy in the performances; any production techniques like double tracking, layering of synths or vocal ghosting and, most of all, the emotion the song is eliciting in them. You can do this kind of listening on a laptop or in your car, hell, take the students to a café and crank a ghetto blaster! Lectures can include listening experiences like these that entice students to think and express what they are hearing, and this imprints domain knowledge of sounds, tones and techniques that they may utilise in their future practice.

Listening sessions are also really important opportunities where students can listen to each other's work, recognise some areas they may have in common and suggestions can ensue to improve the work. Students value these experiences: "Yeah, well the best thing was bringing … in a track, work in progress, playing it to peers, everyone listening with respect … I want feedback for these things and that was excellent" (Georgio, EdD interview 1, February 26–27, 2018). These types of feedback discussions promote networking and future student collaborations, and they can also be used as pedagogical experiences that guide students towards appropriate communication processes. Students will not just instantaneously become good communicators; they need to be educated in how to communicate well and how to give appropriate feedback. Sometimes I physically move this experience to the recording studio (or within tutorials) as can be seen in this feedback session in Video 13 (see Video 13 on the website at www.routledge.com/9780367517632). Again, after an introduction on what the class is covering, students can do most of the talking. This particular video involves third-year students and I use it as an example to first-year students of how their peers have developed their communication practices to that of professionals by third year.

Through the outlines of lectures above, students should develop some integral foundational elements of sensemaking: understanding the different types (scope) of PMP agency and the reasonings to select a certain creative practice.

THE PURPOSE OF TUTORIALS AND LABS: PRACTICE

I find it difficult to imagine a music production culture being replicated in a lab (a room full of computers). Labs are useful for foundational learning (where many students engage in a process on different computers) and perhaps for tasks like individual mixing and programming. However, the social and collaborative elements of music production cultures require people to be working on the same project whilst isolated from one another. Practice-based tutorials help students experience the next sensemaking element: engaging in the basic practices of an activity and then reflecting on it with their peers so that an understanding of practice is solidified. Some tutorial exemplars with which I have had success can be loosely categorised as (1) two-hour learning tutorials and (2) four-hour session tutorials.

Two-hour learning tutorials and four-hour session tutorials

I have provided an extensive list of ideas for tutorial engagements in Appendix B, however, the obvious difference between the two-hour and four-hour tutorial is that some experiences require more time and should embody a PMP session more closely. Two-hour tutorials are excellent pedagogical environments for students to experience a host of the creative practices of PMP, but the musical outcome is often of little importance. For example, the step between writing a song and the manifestation stage (including pre-production) is one of the most confronting periods since it engages with a student's subjectivity gap. To become more confident in this area, students can undertake envisioning sessions where they present original demos (read lyrics or play backing tracks) and the cohort engage with envisioning to suggest prospective ways forward for the production. This is a feedback session that engages with the utility of the producer's vision and students can also use the producer archetypes to inform their discussions as they are all educated in these topics within the lecture series. This process can be seen in action in one of my classes (see Video 14 on the website at www.routledge.com/9780367517632). Other examples of two-hour tutorials could include programming an eight-bar verse and chorus beat for a top-line to be composed over, facilitating a full band pre-production session, double tracking guitars or vocals, layering of synth and drum sounds or the selection and application of effects within the mix of an EDM track (reverbs, delays, chorus, flanger, etc.).

Four-hour sessions can resemble a profession-based engagement where a time limit is enforced for a work's completion. This puts students under pressure in a manner redolent of that which they will experience in the industry. Importance may be placed on the quality of the musical result or the experience can be approached as a learning journey where the result may not matter. In these sessions, students can apply the concepts learnt in the lecture and experienced in two-hour tutorials. Some examples of four-hour sessions with which I have found success include: a complete lead vocal and backing vocal tracking session, sound alike sessions where students try to replicate music of past recordings (Sören Johansson interview, April-May, 2021), a live band recording (see Video 7 on the website at www.routledge.com/9780367517632) or the completion of a mix. I have also used a series of four-hour sessions (say one session a week for five weeks) for students to complete a song from start to finish in class.

The only issue with some of these two and four-hour frameworks is that student class-numbers become problematic in more electronic-oriented genres. It is simply not efficient to have 10–15 students mixing or beat-making compared to the same group undertaking a live acoustic recording with a full band. Flexibility is therefore important; if students can be broken up into smaller groups and sent to various programming suites or smaller studios and your institution has rooms that will facilitate live recordings (basically a room of any size that musicians can setup and play), you can cover most profession-based applications.

THE PURPOSE OF AUTONOMOUS WORK (ENTREPRENEURIAL WORKFLOWS)

Autonomous work provides the student with the opportunity to put into action all that they have discussed and practised in class-based activities. However, autonomous work brings with it an additional element: entrepreneurial workflows. Students should be encouraged to work in an entrepreneurial manner. They should be networking (online), attending social gatherings and performing/running gigs so as to expand their musical social circle. They should seek out musicians that suit their music as opposed to just engaging their friends on their recordings. They can hold creative meetings and start writing and collaborating with musicians from diverse musical styles to expand their musical domain. The front end of the recording (see Chapter 3) must also be considered by students, and appropriate equipment that suits the music should be hunted down, borrowed or hired. This is often problematic because many students have an affection for their own instruments and want to use them continually. Students may need to organise rehearsal and pre-production sessions and ensure the studios they utilise are the most suitable for the music with which they are working. If they can work from home, awesome; if they require the university studios, they should be booking these to complete projects. Students should not work at home exclusively because it is convenient or easy or because they forgot to book the university studios. Often students will use the fact that HE studios are regularly busy or booked out as excuses when laziness or poor time management is the culprit for a lack of studio time. Students should also confront creative conflicts to promote the best possible result for the project and then take time to evaluate their work and get feedback from others that will inform future work.

Autonomous workflows allow the student to employ their knowledge of the greater practice of PMP and fine tune any adjustments to their sensemaking through autonomous engagement and discovery. They should be designing the music production cultures they require for their music and considering the vibe of the people, places and spaces with which they will work. They should be communicating effectively and efficiently with all involved; they can use social media chat rooms, emails or phone calls/texts to do this. Whilst working, they can put into action their knowledge of the three pillars I mentioned earlier (see Chapter 2): surveying of the musical landscape, making music in different ways (diverse music production cultures) and explicit application of music production skills. They can engage with envisioning to target the function of the song and use the producer archetypes as way markers as they traverse music making. Their autonomous workflows afford the student the opportunity to move beyond their subjective bubble and integrate their work with the necessary stimuli from others that will expand its aesthetic or commercial values. The power in this application is that it is the student activating these initiatives, not the educator – these become the building blocks of a professional practitioner. There is no doubting that students may need guidance here and this is where student listening sessions (in class) and educator mentorships come in handy (as will be discussed later).

LECTURES, TUTORIALS AND AUTONOMOUS ENGAGEMENT WORKING TOGETHER

For PMPP I target the concept of sensemaking through an integrated approach of lectures, tutorials and student autonomous work and recommend this process to you. I understand that this approach may be suited to the context in which I teach and may need adaptation for varying contexts around the world. However, it is integral that course convenors understand the overarching ways in which these various frameworks (lectures, tutorials and autonomous work) can work together to increase student learning. Nyssim Lefford exemplified her constructivist approach within her discussions.

> I employ a combination of approaches in typical classes: brief lectures, seminars with a lot of class discussion, group analyses, demos, sometimes workshops where I will visit students as they work, and there is always some project and/or analysis to do on their own (i.e., apply concepts to their own work, and/or "experiment"). It's a very constructivist approach. Every concept is accompanied by an experience of the application of the concept, usually in a situation that allows for experimentation and social interaction. My assignments tend to stress "risk taking". I grade for process and thinking, not for results.
>
> (Nyssim Lefford interview, April-May, 2021)

STUDENT-CENTRED LEARNING

I started teaching PMP in 1998 and for a long time I was troubled with this task due to the sheer amount of knowledge within the domain of PMP. It took a little while for me to realise I was thinking about this problem the wrong way around. Teaching places onus on the educator; learning places onus on the student. I was talking to Jarritt Ahmed Sheel about this and we both thought it was so important that PMPP nurtures student-centred approaches:

> We were talking about what kind of experiences like each program can offer students that they can't get out in the real world. It has to be student-centred learning. Like, "what do you want to do?" See that's the problem with the conservatoire method … it is so canonised and about "how to do something right" and about the literature. Instead of asking the student "what would you like to do?" Meeting the student where they are. That's what we need to do.
>
> (Jarritt Ahmed Sheel interview, April-May, 2021)

Student-centred approaches, in my opinion, are amongst the principal practices that institutions can use to facilitate an education in PMP. Focussing on this practice obviates the need to exemplify every production technique that has occurred from the 1920s to now. Once some foundational knowledge of the practice of PMP has been translated then communication can ensue between educator and student to locate the student's musical aspirations. Educators, at this point, can offer some advice on possible influences or processes that may be useful. Student engagement then shifts the education experience to something individualised for that student but contextualised within the overarching context of PMPP. Bryden Stillie and Zack Moir elaborate:

BRYDEN: I think over the last maybe five years I've definitely started to see the change in the way our degree structure works so that it allows people to become individuals and understand more about where they fit within the world and what they might be able to achieve, rather than pigeon holing them.

ZACK: I agree, I think it's about making the motivations intrinsic, I think it's about making the scope global, so it's about saying right okay here's a structure, a kind of shell, you can do whatever you want within that shell. So then we get questions like, why are we doing this? Why is this important? Well let's turn it around and say why do you think it's important? What do you want to get out of this? So I guess it's about making the experience as customisable as possible, which is kind of about the student centeredness, but it's also about flipping it so that they're not doing an assessment to pass so that we can validate what we're doing, but they're doing an assessment as a ladder to climb as part of their learning. It's about scaffolding, but more importantly it's about building shells and building safe spaces for them to work within, such the motivation becomes intrinsic.

(Bryden Stillie, Zack Moir interview, April-May, 2021)

Student-led research: Informed practice

Student-centred approaches can be aided by the integration of student-led research practices. These can be ad-hoc or integrated within assessment items. Encouraging students to undertake some musicological research into artist's seminal albums, production techniques or socio-cultural influences on music can be influential in the way students consider their production activities. The key here is letting students choose who, or what, they may want to research. Research can also aid in the facilitation of a well-rounded education that may culminate into postgraduate study. I spoke with Mads Walther Hanssen about the utility in engaging students with musicological research and how, by developing thinkers, institutions may promote innovative graduates:

> For me, music production is a part of what I would call musicology. There is no distinction... In Denmark, the conservatoires are practitioners and the universities are thinkers... I think the world needs more thinkers. There are so many good producers out there, thinkers are people who can develop something new in the future.
>
> (Mads Walther Hanssen interview, April-May, 2021)

SUMMARY

I advocate that educators position their pedagogical frameworks within a holistic goal: sensemaking of the greater practice of PMP. To do so, they can then combine knowledge acquisition through lectures, listening sessions and practice-based applications that are accompanied by reflective discussions and student-led research. These concepts can be both holistic to the group and/or dialled down to student-centred approaches – every student is developing an individualised understanding of the practice of PMP. In my opinion, the greatest challenge educators have in this regard is to ensure the educational experience embraces popular music culture and attempts to capture and simulate the variables this culture introduces. The majority of student engagements should be socially interactive; students and educators should, whenever possible, be encouraged to embody the actions of popular musicians so that rich, entertaining and fun experiences proliferate. By doing so, institutions are normalising these activities for students and preparing them for the many socio-musical interactions they may encounter in the future.

REFERENCES

Association Européenne des Conservatoires. (2017). AEC *Learning outcomes 2017*. Retrieved from https://aec-music.eu/publication/aec-learning-outcomes-2017-en/

Pensado, D. (Producer). (2013). Mix engineer Chris Lord-Alge. Pensado's Place #80. Retrieved from http://www.youtube.com/watch?v=pqZHOrj1VjQ

Quintana, C., & Fishman, B. J. (2006). *Supporting science learning and teaching with software-based scaffolding.* In Annual conference of AERA, San Francisco, CA.

Voss, Brett (2018). *'The design principles of on-demand learning: A design-based research study of educative provisioning in popular music production'* (Doctoral thesis, Griffith University, Australia).

12

The music

The popular music with which students work in PMPP should be appropriate for the learning outcomes of the program, but it should also resonate with the student's intrinsic desires and aspirations as musicians. As pedagogues, if we are to tailor our programs to student-centred approaches (a tactic for which I advocate), we should seriously consider the value of students working with music that means something to them. This is a complex problem because, as I will exemplify throughout this book, there is merit in both teacher and student-informed work and the student experience can be negatively impacted if students are working with music that they feel is inappropriate. International student survey responses exemplify this "[we never worked on] techno or house which are my style. Just simple popular genres" and "the professors gave us a list of classic songs to cover. Original music was not permitted for one class. Another mixing class allowed original music but it couldn't be your own. This was disappointing" (International student survey response, May, 2021). So, what type of music is most useful in PMPP settings? Perhaps in these instances it is crucial to understand the utility of the music we use in our pedagogy. What is its purpose?

AUTHENTIC MUSIC

Within more advanced PMP contexts, students should be involved in practices that promote the generation of authentic music. Authentic popular music is defined here as something that is genuine to the individual who created it, believable and something that may be simultaneously relatable to an explicit model and original (Peterson, 1997, p. 220). However, authenticity in music is often validated through perception (Moore, 2002) and "requires a reliance on tacit understanding" (Parkinson & Smith, 2015, p. 94). Moore (2002) suggests that authenticity should "shift from consideration of … originators towards … perceivers, and should focus on the reasons they might have for finding, or failing to find, a particular performance authentic" (pp. 220–221). Therefore, when discussing authentic PMP, it may be more beneficial to consider who the popular musician is and the context in which they are being authenticated (Moore, 2002). For

DOI: 10.4324/9781003055099-16

the purpose of this book, authenticity is suggested to be a construct that is intrinsically linked to the musical world in which an individual operates (Smith, 2002, as cited in McIntyre, 2012, p. 162). This definition considers the production of both original and non-original compositions as potentially authentic.

Original material and covers

Original material requires the producer to consider a piece of music that has never been recorded before. This adds complexity to the music production process because it challenges the producer to interpret musical and sonic semiotic representations of a song's meaning, design new listening expectations for the listener and this is a different engagement of music production than producing covers alone. The production of original music is imperative for students of PMP to experience to aid their industry integration and provide diverse revenue streams that will promote their longevity as producers.

Many of my students come to university thinking that their original music will be industry defining and pave the way to a fruitful career. The reality is far removed from this; when most students, even the most talented, come to university they are often one-dimensional, and this is simply because they have not ventured outside of their "subjective bubble" (Nyssim Lefford interview, April-May, 2021). To combat this, students can be empowered to work with music in diverse ways. Collaboration has utility here, especially between students who may operate in disparate musical worlds: Heavy Metal and Pop, for example. Covers are also useful to push boundaries because students have less emotional attachment to songs they have not written and are therefore happy to explore different production styles using them as the foundation. Often, I will ask students to pick a song from another artist who works in a similar style to them and re-imagine that song in a contrasting style. Or alternatively, I will ask students to contextualise music from a contrasting genre into their current style. Either way, students are working towards the generation of authentic music. Student survey responses supported the use of genre/blurring to diversify their creative practice and broaden the scope of their musical works. For example:

> I liked working in my preferred genre.... On the other hand ... when we were introduced [to] cross-over genres or other exciting subjects, I often found myself using this inspiration to explore new [ideas]-leading me to broaden my perspectives [and apply ideas of other] genres in my own original music.
>
> (International student survey response, May, 2021)

A classic example of a song that has been re-imagined in this way is the unplugged cover of "Take On Me" by 1980s pop band A-ha (Furuholmen et al., 1984). When I play these two contrasting versions in class, students are always amazed how a completely different emotional response is successfully evoked in the new version. "Take On Me" was a massive hit in the 1980s and this new version tailors to the consumer in a clever way. Many people who were listening to music in the 1980s (like me) loved the up-tempo synth-pop style of "Take On Me". However, now that this audience has matured (some 36 years in my case), the unplugged version allows them to re-connect to teenage memories in a melancholy way. The use of old instruments in the remake connects the listener to historical concepts, the slow tempo and lyrical messaging emotes a nostalgic feeling of longing and searching for love, or perhaps one's youth. By targeting this nostalgic feeling, A-ha's pop hit from the 1980s has been reimagined to target the same end listener but in a new way.

Commercial or self-expression: The song's function

PMP pedagogical approaches are much more complex when dealing with original material. This is because students need to understand the realities of their songwriting approaches: whether they target commercial goals or emphasise self-expression. Joe Bennett discusses this:

> The word 'self-expression' brings up the issue of authenticity in pop. If you're a commercial pop songwriter no one gives a damn about your personal self-expression – that's not your function. Like most adults, you've probably been in love and had your heart broken, you've probably been dancing, and you've probably asserted your self confidence in some way. These are all part of the human experience, and I suggest the main subject matter for songs. If an arts-based degree wants to encourage artistic-self-expression, but teach commercial norms, then the teaching of songwriting creates a natural tension, because commercial popular music operates within known artistic constraints. In an all-expressive course, the students can write any kind of song they like, and it doesn't matter if it meets commercial standards; the outcome is just to develop oneself as an expressive artist. But if you are saying "we are a music degree and you are going to become a music industry professional at the end of it", and part of the activity is writing songs, then that self-expression has to be subservient to commercial activity, at least to some extent.
>
> (Joe Bennett interview, April-May, 2021)

Joe went on to suggest that educators should look at the audience, and how music is crafted and use that to inform curriculum design. He suggested this was to engender respect for the audience as opposed to cheapening the value of creativity.

> A playwright will make edits to a scene to make its important points land effectively, and within the run time of the theatre show. Well, I suggest a songwriter needs to do exactly the same, which means I think, for us as curriculum designers, we need to put aside any romantic notions of autobiographical self-expression in song writing. It's just love songs most of the time, and it's, as a colleague said to me once, there are only

> spring, summer, autumn and winter love lyrics. You know, looking for love, being in love, love on the rocks and love lost. The four stages of love that we have as humans, and obviously from the ancient Greeks to the present day, poets and artists and creators of all types have found endless ways to tell that story.
>
> (Joe Bennett interview, April-May, 2021)

Often students will reply to these discussions with disdain for the commercialised nature of some songs and will insist that their music will be industry-defining:

> I hear that from students all the time, "I don't want to be pandering to what's going on now, I'm going to be ground-breaking". Well, pop music does innovate; pop music has a massive track record of innovating. But I suggest that its innovation has been sonic, technological, harmonic, and structural – almost any parameter other than lyric subject. Lyric subject is the one thing that is remarkably consistent. Sometimes a particular lyric theme may become popular in a specific niche of popular music; take Gangsta Rap as an example. There was no equivalent of that self-aggrandising, socially violent lyric in 60's rock for example. Gangsta Rap is its own thing: it is lyric expression within a tight set of parameters, and if you want your students to write in that style, then studying that repertoire and understanding how it communicates is important. But I think when a student comes up with an argument that they want to write 'ground-breaking lyrics', and I've heard it literally hundreds of times, I think we should explore gently and sensitively, as curriculum designers, what's behind that – why are they saying that? They must know that for tens of thousands of years people have been writing love songs.
>
> (Joe Bennett interview, April-May, 2021)

The disparity that Joe discusses between students' expressive aspirations and the commercial realities of the music industry has been prevalent in my students. When I questioned Nyssim Lefford about the difficulties we face in PMP, she addressed some integral considerations of commercial product and/or art form:

> The very notion of "popularity" can divide teachers and students. The teachers learned at a time when there was less music produced. The content that was released passed through gatekeepers, critics and tastemakers. All these layers, shaped popular tastes and consumption.
>
> (Nyssim Lefford questionnaire response April-May, 2021)

In contrast the Internet has catalysed a situation where students may feel comfortable working in their style (appealing to niche audiences) and subsequently steer clear of producing music outside their comfort zone:

> Technology has made production easy (easier) and comfortable. We now have [technical] means to verify conformity or "correctness".

> At the same time, the classic, historical recordings that teachers regularly use to illustrate concepts and possibilities were made from blood, sweat and tears. The best of these examples are the ones that broke rules, went against the grain and set new standards. They were essentially uncomfortable (at the time of their release). How can teachers reconcile these differences while discussing the rationale for production decisions? Ultimately, there is nothing comfortable about art – for either the artist or the audience. Music productions, like other modes of artistic expression, are meant to move listeners out of their emotional and intellectual comfort zones. If students are uninterested in exploring difficult territory, how can we teach them the art of production?
>
> (Nyssim Lefford questionnaire response April-May, 2021)

Nyssim's and Joe's words highlight the complexity of this situation and allude to how important rich and meaningful discussions between the educator and students can help both parties unpack why students feel a certain way. Empowerment from the educator in these situations may encourage students to contemplate the diversification of their creative practice and their futures as working professionals and/or expressive artists. Subsequently, I understand why varying programs consider pedagogy from commercialised or art-centred contexts. However, it is my opinion that these are choices for the students of PMP to make, not the institutions; it is not in me to limit students to solely writing and producing songs that target current understandings of commercialism, for example. My students have suggested to me that higher education is a safe place for them to explore their expressive intuitions without the overlying pressures of income generation. Higher education can facilitate opportunities for students to work out who they are as creative musicians, and these activities can take time to thrive and become insightful. I also do not feel there is value in empowering students to produce music that has limited utility in the commercial world by always placing emphasis on self-expression. For music education to be equitable, diverse and inclusive, the aspirations of the students need to be nurtured. My conversations with Jarritt Ahmed Sheel unpacked how many PMP graduates will go on to use music as an avocation, something that is not their main source of income but something that they hold very dear to them. Therefore, to accommodate both commercialised and expressive intentions educators can have open discussions with students and steer their attention towards recognising the *function* of the song: what is the song meant to accomplish? Self-expression? Commercial success? Something that includes both of these in an innovative way? (Sören Johansson interview, April-May 2021).

By developing professional relationships, mentoring and engaging in conversations with students, educators can help them to critically consider the realities of their musical decisions. This is what best represents my own intentions founding my pedagogy: providing an enlightening experience that illuminates the many pathways that are possible so that students can focus their futures on that which they feel is most relevant. For some, the problematic element in this approach is grading a self-expressive work. Commercial works can be measured against current hits. Conversely, a student's art is more abstract, and many educators struggle with critiquing original material.

> Teachers are so scared to tell a student if something is bad that everything is glorified and magnificent. That's why original compositions don't work at our school. One teacher out of the whole curriculum was able to actually roast students for their work, and to me that is the only way to really teach someone. Tell them the honest truth of where they are and how they're missing.
>
> (International student survey response, May, 2021)

Many people view expressive works and songwriting as subjective and therefore difficult to critique within public forums. Subsequently, some educators can become hesitant to open songs up to feedback and, for this reason, steer clear of original works. I believe the fact that songwriting is subjective is the quality which makes it open for critique. Music is relentlessly critiqued as a part of the listening experience and when it is consumed. In PMPP, the way a song can be equitably critiqued is to establish the function of that song. It is then just a matter of considering if the song has accomplished the writer's intentions to the subjective estimation of the listener. Whether a song is written for self-expression or consumerism it is rarely created in a vacuum. Most artists aspire to having people listen and experience their music in some way. Therefore, the function of the listening experience can always be unpacked and critiqued, and this then becomes a viable framework for working with, and critiquing original material in HE.

PROMOTING GENRE AGNOSTIC PRACTICES: ASPIRATIONS VERSUS INSPIRATION

Locating pedagogical approaches within specified types of music (electronic music, for example) helps to frame appropriate learning outcomes and position students inside a domain of practice. There is utility in this approach should various institutions wish to attract students who prefer these types of music. However, at the same time, these approaches have the potential to constrain students within an unrealistic form of professional practice, after all, genre blurring is rife in the industry today (Paul Thompson; Nyssim Lefford interview, April-May, 2021). In PMPP, genre is currently more useful if considered as a form of typology (an understanding of a concept) as opposed to a constraint of practice. Genre codes are things that practitioners consider when working: sometimes they can be followed, other times they can be contextualised with other genre characteristics, and often they can be discarded. Musical boundaries are consistently evolving; in the 1980s the use of synthesisers and the saxophone was prolific, so much so that in following decades (1990s and 2000s) they were often frowned upon as instruments of choice. This was quite depressing for myself seeing as I played both keys and saxophone in the 1980s! However, there has been a resurgence in these musical instruments in various musical styles, including EDM. For example, "One Day" by producer Bakermat (King Jr, &

Lodewijk, 2012), integrates a sample of the "I Have a Dream" speech by Martin Luther King Jr with a saxophone melody, and this track has reached approximately 31 million views on YouTube alone.

Considering genre as a fluid typology allows educators to locate pedagogical frameworks within students' aspirations (electronic music or rock music, for example), without limiting a student's inspirations: all past musical styles. Why limit the boundaries with which our students may choose to seek inspiration? Why shouldn't Hip Hop producers be knowledgeable about the acoustic musical styles of the 1960s if these ideas, when fused with the students', produce an authentic work? Educators do not have to teach everything; students are more than capable of doing autonomous research to unpack the intricacies of a certain style of music. By opening curricula to the vast scope of music possibilities, a student's domain of knowledge (inspiration) that can inform that which is possible (aspiration) can become expansive. Students from an international survey valued genre agnostic practices

> Most of the assignments were kind of free genre-wise, which was cool (ex. make a cover of a Kate Bush song – in your style) and then I could choose to work in my genre (acoustic, experimental singer/songwriter). But sometimes it was more like "create a song that Ed Sheeran would perform" or "do something for the Eurovision Song Contest" and then you had to go out of your own genre which was always a good challenge to consider music in new ways.
>
> (International student survey response, May, 2021)

With genre becoming less of a constraint to PMP agency, there is the potential for the music that producers use as influence to bleed across past traditions. In the future, popular music, jazz, classical and music theatre, for example, could be fused together in new production amalgams. As such, understandings of what constitutes popular music will undoubtably remain in a state of transition. A key takeaway here is that students can be empowered to apply critical enquiry and innovative applications in the creation of new musical styles. To do this, students can use both the traditions of current pop music but also draw on new territories, act intuitively, fuse electronic and acoustic styles should they choose, but, most importantly, become flexible music makers.

PROMOTING STUDENT FLEXIBILITY: ACOUSTIC, ELECTRONIC AND ONLINE MUSIC PRODUCTION

I was speaking to Heather Kirby, and we agreed that student collaboration within diverse music-making environments was important so that graduates developed into flexible practitioners:

> I find a variety being useful … [getting students to collaborate in home studios, online and large studios] helps to instil a different kind of confidence in students…. Like "I can do this at home and in [large

> studios]" and this helps … students to be flexible and adaptable in different social situations…. Students come in wanting a certain thing [collaborating with others], others want to make music on their own. So there is a wide variety needed so that each student gets a lot out of the diverse and valid ways to think about music production.
>
> (Heather Kirby interview, April-May, 2021)

Heather and I went on to discuss that it was important to place students in both comfortable and foreign collaborative frameworks because this helps develop practitioners who are adaptable to different types of music, client personalities, workflows and demands. The COVID-19 pandemic also brought with it limitations in PMP professional practice, and flexible practitioners (of which Heather and I know several) continued working though the pandemic because they had the ability to collaborate with many styles of music and in virtual and online music-making scenarios.

In my opinion, we must entice all students to obtain a holistic understanding of music making in electronic, acoustic and online platforms no matter the foci of the program in question. I do support programs that bias their approaches 60% electronic, 20% acoustic and 20% online, for example, but this is more difficult than it seems because students are so focussed on their own musical interests. However, the inclusion of teacher-informed works that target the diversification of students' skills in a wide range of music-making scenarios will support a graduate's future as a professional practitioner. This is where I believe that foundational competency in the three technospheres (electronic, acoustic and online) is essential to all students.

The application of sound engineering, critical listening and programming skills are constant in these three environments, but session management, songwriting and the ways in which producers communicate and collaborate in online environments are unique and often problematic if experienced for the first time:

> [You miss] all the interpersonal interactions happening in real time. You know when you have a group of people in a room the complexities of relationships are a huge factor. You don't necessarily have an opportunity for such dynamics to come into play online. Having said that maybe it's good that we can eliminate some of the distractions … I do think most people miss face-to-face collaboration, but what's a little less clear is whether they miss the familiarity or they're just not comfortable, they like being a social animal. Or maybe they just resist change, they want to do things the way they know. Ultimately, we may be surprised to find out that remote collaboration has its own charms and perhaps is equally valid.
>
> (Andy Edelstein interview, April-May, 2021)

There are other lessons to be learnt within online music collaboration. For example: the sharing of files (stems and sessions), cross platform DAW operation (Logic Pro and Ableton Live, for example) and the pros and cons of synchronous and asynchronous collaboration. I engage my second-year students in an online collaboration project with students from Edinburgh Napier university. This initiative, that was developed in collaboration with Zack Moir, forms a part of student course work and assessment. The process afforded students the opportunity to write and produce original material with people from diverse cultures and via online and remote means. It has proven to be a great learning journey for students concerning communication and organisational workflows as was suggested in student's reflective reports and self-assessment results. As educators, we must work together and engage students with online music-making experiences as more and more possibilities for this type of collaboration are developed (Avid Cloud Collaboration, for example), and larger quantities of music are produced this way.

SUMMARY

My goal in this chapter was to outline the role of the music in PMPP and how it can positively impact the student experience and their development as popular music practitioners. I do not think there is one model that will fit the diversity of programs around the globe other than to suggest a large proportion of the students' music should target authentic music production, should include original material, resonate with the student's aspirations and be inclusive of acoustic, electronic and online applications.

REFERENCES

Furuholmen, M., Harket, M., & Waaktaar, P. (1984). Single (From the Album: Hunting High and Low): Warner Bros. Retrieved from https://www.youtube.com/watch?v=-xKM3mGt2pE

King Jr, M. L., & Lodewijk, F. (2012). (Produced by Bakermat). Single: Délicieuse Records. Retrieved from https://www.youtube.com/watch?v=6BV37Dcxj9s

McIntyre, P. (2012). *Creativity and cultural production: Issues for media practice.* London: Palgrave MacMillan.

Moore, A. (2002). Authenticity as authentication. *Popular Music, 21*(2), 209–223. DOI: 10.1017/S0261143002002131

Parkinson, T., & Smith, G. D. (2015). Towards an epistemology of authenticity in higher popular music education. *Action, Criticism, and Theory for Music Education, 14*(1), 93–127. Retrieved from http://act.maydaygroup.org/articles/ParkinsonSmith14_1.pdf

Peterson, R. A. (1997). *Creating country music: Fabricating authenticity.* Chicago, IL: University of Chicago Press.

Smith, G. (2002). *Response to the Keynote Address by James Akenson.* The First Annual Australian Institute of Country Music Conference, Gympie, Qld, August 23–25.

13

The educator's role

When I contemplated writing this chapter, I decided it was important to be perceived as authentic as possible because this is what I embody with my students so as to build relationships and mentorships with them. The university culture is an intriguing place for this to occur. I remember watching the movie *Patch Adams* (Shadyac, 1998). In this movie, there is an iconic scene where the dean of the university is trying to convince members of the education board that Patch Adams should not achieve his doctoral qualifications due to his "excessive happiness". There is a constant elitist underpinning throughout most of the movie that medicine is a serious business and therefore the art of doctoring should be as well. I know that if you have watched the film and investigated the very real journey of the eponymous physician, Hunter Doherty "Patch" Adams, you will understand the incredible accomplishments of his work with people. Unfortunately, some responses from an international student survey I facilitated suggested that their PMP educators were "elitist", "always talking about themselves" and "wanted us to learn their way" (International student survey response, May, 2021). PMP should be a fun experience; learning cultures should steer away from traits like elitism and instead replicate the enjoyable social realities of making popular music. The education of PMP must connect to that which is successful in the field. Therefore, I propose to you that no matter what framework you put in place or how stringently you select your students, if the people on the ground who are facilitating the role of educator are approaching the pedagogy in an unsuitable way, then the possibility for high-quality learning to occur is minimal.

WHO IS THE EDUCATOR: INDUSTRY PROFESSIONAL OR ACADEMIC?

If the programs we provide are intended to develop graduates who are innovative practitioners capable of critical enquiry, then ideally PMP educators need to have both experience in the industry as working professionals and academic qualifications/experience in PMP research. This may seem like a *perfect world* type scenario but, in fact, I believe it should be a pre-requisite, and staff should be hired under (or with the

DOI: 10.4324/9781003055099-17

inclusion of action regarding) these criteria. My discussions with Simon Zagorski-Thomas alluded to the fact that HE has been doing "bad education" whilst the industry transitions from apprenticeship models and producing university graduates who can't walk into studios and produce musical works. If these people go on to be educators (because they wont be employed as practitioners), then there are obvious limitations in this model. Additionally, an educator's industry experience should include modern PMP and be worldly, or at least not located explicitly within the town they originated, studied in or made music in as this can produce educators with a narrow perspective of the industry. Students expressed concern regarding the suitability of some of their educators:

> Most all teachers were hired because of their previous experience, so naturally my university floods the school with professors over the age of 50 (of high regard), who know how to operate an analog console and do things like they did in the 70's and 80's. That is why my school relies on analog consoles for the majority of audio engineering. Most of the professors can teach the science, but when it comes to modern music production, not so much. My college is easily behind the curve because they don't hire for the future, they hire because of clout and friendships or a professor knew a guy (imagine millions of dollars entrusted to a department that thrives off of the boy's club). When I sit in on the new professors, it's more of the same. None of these professors have even opened Ableton or Serum or used a sub37 keyscape or omnisphere. No professors built a studio in 2020 with their bare hands and made a hybrid studio of analog and digital concepts.... I'm just annoyed that some colleges are going to be swimming in the past for a long time until their professors retire, and they get professors that are with the times.
>
> (International student survey response, May, 2021)

I also believe that, regardless of experience, any industry practitioners who transition to educator roles and downplay the value of postgraduate qualifications or research–led approaches in practice and pedagogy because they "learnt it all out in the real world" should also be shown the door. In my early days as an educator, this is what I thought, and I was wrong. Many professional practitioners know how *they* needed to work to get things done, but they have not learnt the value in being flexible pedagogues to promote student-centred approaches. "A person entering the production field who only learns from a working professional may get a valuable but narrow perspective when it comes to learning essential skills" (Burgess as cited in Frith & Zagorski-Thomas, 2012, p. 269). Therefore, educators who have no idea about experiential, situated or phenomenological learning frameworks have no business convening PMP courses. We must explicitly consider who our educators are because our goal should be to produce students who are capable of critical enquiry into the development of *their* best practice; they should not be attempting to replicate the practice of the educator. These traits will upskill graduates for their futures as professionals because they will be able to reflect on their work and adjust their approaches as the industry evolves. However, we cannot ignore that students relate well to the fact their educator is an industry professional because this helps the students feel connected to the working world. I discussed with my students why they valued my experience as a professional producer:

THOM: Yeah, 'cause you've had first-hand experience in all that.
RESEARCHER: But I don't talk about me very much, do I?
THOM: No. But I guess you're known for doing that stuff, so people take it more seriously like, "okay this guy does know what he's talking about".

(Thom, EdD interview 3, July 9–10, 2018)

I always find it interesting how students find out about an educator's experience. I have listened to many educators verbalise credit lists to students so as to gain their respect. My approach is much more passive, because, let's be honest, no-one likes a name dropper. I leave my website address in plain view on emails and class online platforms and student research and word of mouth does the rest. The rest of the time I am just me. I suggest you do the same.

Students also like it when their educators reflect on professional situations that they have been involved in. As an educator, I have learned that these types of conversations help to build self-belief in students by nurturing the educator/student dynamic. However, these should be very occasional, relaxed informal discussions, and they should be used to convince the students that they have a lot in common with their educator and, therefore, can similarly become professional practitioners. An example of one of these discrete conversations can be viewed on Video 15 P. Some of the students and I were having a pre-class discussion (regarding alcohol), and this is one instance where I did discuss my personal journey as a professional to help the students understand some of the realities of PMP (See Video 15 on the website at www.routledge.com/9780367517632).

EXEMPLIFYING PRACTICE VERSUS FACILITATION AND STUDENT-LED APPROACHES

There is value in both showing students (exemplifying) how a certain practice is completed and allowing students to experience the doing themselves. I discussed this with Prince Charles Alexander and the wonderful video resources he has put together where he is exemplifying practice and talking about how to produce music. An example of this work is the iZotope (2018) videos.

PRINCE CHARLES: … I actually produce somebody. I always bring a singer in and [students] see me do that with someone [produce a lead vocal]. Because I saw it done as an assistant engineer … I saw some producers do it and that's how I learned how to do it.
BRENDAN: Do you ever go "okay, you guys have a go at that?"

PRINCE CHARLES: I feel they learn more from seeing me do it because [I am using] my lens and I think that each student has to discover their lens and discover their comfort level of how they have these communications.

(Prince Charles Alexander interview, April-May, 2021)

Prince Charles went on to discuss the necessary conversations educators and students can have when listening to music and how mentorship (of the student lens) can occur in one-on-one or small group sessions. These discussions can outline musical targets for students during autonomous work since, ultimately, students experience the realities of their practice whilst working outside of class. Interestingly, the students of my advanced PMP courses appreciated me exemplifying techniques to them but clearly preferred when they were able to work alongside myself, as musical peers, in tutorials. In one of my final doctoral study tutorials (See Video 16 on the website at www.routledge.com/9780367517632), students took control of a backing vocal session and I was relegated to just singing backing vocals. This was the students' last class for their degree, so there was a jovial attitude. However, you can see that our relationship had developed to a location where they felt comfortable critiquing my singing (they refer to me by my nickname, "Killer") and having the confidence to direct me like a backing vocalist. They even used humour to do so, which I thought was exceptional (see Video 16, 3:00–3:17 and 4:50–5:10). In this instance, students were doing the work and learning at the same time. This model is one where the educator is more of a facilitator, as Jarrit Ahmed Sheel explains:

> So the educator's role should be that of a facilitator, that's it. To be a good and kind guide of learning. So a guide doesn't force you to go a certain way. They go "oh look what's over here". Some teachers have been led to believe that guiding someone is forcing them instead of letting things wash over them, facilitating an experience. Let it be about student-centered learning. Bring them some things of where you want them to go so they know where to visit when they're ready.
>
> (Jarritt Ahmed Sheel interview, April-May, 2021)

Cremata (2017) highlights some educator approaches that promote the practice of facilitation:

> A popular music facilitator, responding to his/her students' needs, regulates control levels and differentiates instruction by giving and removing assistance. Rather than focusing on blend, balance and uniformity (aesthetic qualities), a facilitator emphasizes individuality, differentiation and freedom (social qualities). This is a shift in values and in the sorts of habits traditionally promoted in music education contexts. This research supports the notion that facilitation in popular music education contexts promotes democratic, autonomous, diverse, differentiated, collaborative and inclusive learning environments.
>
> (p. 76)

Cremata's words above reinforce the concept that PMPP should promote intuitive and instinctive pedagogical delivery that engages student individuality. Facilitation promotes students driving the learning process where learning becomes tacit while they make music. This not only bridges the classroom experience to the profession and feels real to students, it also removes any reliance on the educator to provide education into the future. Resultantly, students take control of their learning, and this is imperative in order for them to remain connected to an ever-changing professional world. Once foundational knowledge has occurred and important PMP practices exemplified, an educator's engagement within advanced classroom activities could embody the following:

> My educator was a mix of all of these (teacher, industry professional, facilitator/ mentor). He engaged with us as peers and adults in order to communicate with us effectively. In most practical classes, he would simply be in the room, casually engaging with the students and facilitating the session as needed. When the class was noticeably losing focus, he would step in and talk to us about it, pointing out where we were going wrong and providing guidance as to how to regain our direction. I personally found this approach to be highly effective, as it removed the teacher/student dynamic that many other learning environments have, and made the experience so much more personable and enjoyable for me. It also allowed me to have much more hands-on experience and control during classes, which greatly improved my music production skills.
>
> (International student survey response, May, 2021)

A MUSIC PRODUCTION PROFESSIONAL, COACH, PEER, THERAPIST?

I don't know if it is simply in my DNA as a professional producer of 33 years' experience, but my social interactions with staff and students are reminiscent of how a producer acts. I am forever developing authentic temporary relationships with staff and students so I can build the trust needed to work with them on many different levels. Prince Charles Alexander agreed: "You know, I feel like I am part coach, part therapist, um … I'm trying to make it okay for them to say anything they may need. 'Cause, you know, I'm at a pretty intense music school …" (Prince Charles Alexander interview, April-May, 2021). Prince Charles went on to describe the incredible skill level of jazz musicians at Berklee College of Music, differentiating the school's music production students who need to be okay with thinking of simplicity as an advanced concept. To help students feel comfortable with these challenging experiences, in-depth personalised discussions often help pave the way for student confidence and clarity. In these situations, students value communication with an authentic human being:

> I mean, personally, I think you should be your authentic self because nobody is wanting to learn with a person they think is a jerk.... I ask

> this all the time, "whose ever had a real or like grand learning experience from a teacher they thought was an absolute jerk?".... You know? So, if you're yourself, it doesn't mean you have to be real close friends with the students, but at least you can communicate [on a similar level with them]. That's like being in the studio with a real producer.
>
> (Jarritt Ahmed Sheel interview, April-May, 2021)

In the end, I believe the social landscape of PMPP should be just like making music in the studio. If educators are authentic and can gain their students' trust through in-depth discussions on many levels, then student self-knowledge and personal development can be targeted thereby facilitating the realisation of future aspirations. My discussion with Alexis Kallio uncovered a commonality in our approach which links to authenticity:

> Equal power relations are maybe an impossible destination, but I think that we can work towards such a democratic educational environment when the teacher is learning as well. Do the same reflexive work, the same co-learning and exploring. I heard a colleague say that he 'teaches on the edge of his expertise'. I like that - teaching on the edge of what we understand and always learning (with the students) together.
>
> (Alexis Kallio interview, April-May, 2021)

RELEASING CONTROL

My research has found that, within advanced PMPP processes, if educators aspire to promote graduates who will become innovative practitioners, they should often release control of the educational experience when the students are ready. Simon Zargorski-Thomas discussed this with me:

> Just listening ... and not taking control and ownership over things [is so important].... It's about asking the student "what feels authentic or right to you in terms of your performance practice, your musical sound?".... I think what I'm guilty of is expecting them to be more mature than they are, because I'm mostly teaching postgrads. Having a daughter going into HE now reminds me of how much they still basically are children with a bit of freedom. You can't just flick a switch and say you're responsible now. That is something that has to be a big part of that three-year process.
>
> (Simon Zagorksi-Thomas interview, April-May, 2021)

Zack Moir also had ideas regarding this:

> So, if what we're doing is giving them opportunities to be themselves, to critically look at the world around them and see how they can fit into it or change it for the better.... Then we should be doing all the things we are talking about. We should be having genuine realistic relationships with them. We should be helping them learn the things that they feel they want to learn, not telling them what the

> content is. We should be showing them the boundaries of fields and saying "you could push through these if you want to". I think music is the perfect place to do that to be honest. Because it is so intrinsically linked to culture and to personality and I think that all the things we talk about that we're trying to do … are steps towards that. But you can't just do that without explaining why you're doing that. So I think the important thing that's often missed in music degrees is linking the strategy to the problems that we see in society that we're trying to overcome.
>
> (Zack Moir interview, April-May, 2021)

Education that exemplifies Zack's words requires the facilitator to have a progressive mindset, one that does not look to historical frameworks that covet control (master-apprentice models) but instead considers the sociocultural aspirations of the modern student. As Zack has alluded, students should be made aware of the importance of linking their strategies "to the problems that we see in society that we're trying to overcome". The students of today are the industry innovators of the future and educators must allow them the freedom to consider solutions for the current inadequacies of the music industry. Students should also be allowed to take control of the learning experience and apply that which *they* feel is suitable, reflect on their outcomes and learn from their shortcomings. This process develops confidence in students' mindsets and facilitates the development of innovative applications. However, even though the facilitator may have released control, students will still need support. Mentorships (which I will discuss in a separate chapter) are always helpful in these contexts where meaningful conversations can take place at any time to discuss music life and inspire students' thoughts and ideas regarding their agency.

Disappearing act: Student ownership

To promote students taking control of the pedagogical experience (when they are ready) I often intentionally leave the tutorial room whilst students are working (See Video 2 on the website at www.routledge.com/9780367517632, 5:33–7:57). In my research, the students suggested that they just kept on working in my absence, and that there was a change of comfortability because my leaving also removed the concern of "what's [the educator] thinking?" (Anthony, 2019). Removing the educator at various times from the tutorial experience not only siphoned out the students who really did not want to engage, it also encouraged the students to take ownership of the tutorial and made them follow their intuitions which, in turn, promoted their self-belief as professional practitioners. This is a useful tactic for educators to use and then reflect upon with students as a learning outcome. I questioned students about this pedagogical approach and it was highlighted that this should only take place once students are comfortable in their music making environment.

HAYLEY: We saw you leaving but we were sort of just like, "oh, BA's walking out, he'll be back" and we just kept doing what we were doing, like …
THOM: That was awesome. We felt like we were making gold!
HAYLEY: Yeah, [but] if you did that at the start, it wouldn't work. We would've been so lost.

(EdD Focus Group 1, May 29, 2018)

HONESTY, WHO'S THE BOSS? REALITY CHECKS FOR STUDENT ENGAGEMENT

There is nothing more valuable to the educator/student relationship than brutal but well-considered honesty. This is especially appropriate when discussing student work; students respect educators who articulate areas that need improvement and subsequently offer advice as to how such improvement may be implemented. However, there is danger in over nurturing students; sometimes students need to feel the realities of their actions. In these cases, educators need to consider how much they should influence the educational experience, and this is often an intuitive process that is different with every class iteration. My doctoral students discussed this in a focus group:

ELLIOT: Well, like the level … of how much BA [educator/researcher] like would engage … I thought you actually engaged too much … but I thought it got better over time.
LENNON: But see I disagree because in the beginning, I think we needed more of that engagement, so we could learn … and then once BA kind of was in tune with us and … we'd learned a bit more, he backed off a little bit and then it became more of a self-led [process].
HAYLEY: When we weren't stepping up … you assisted us but when we were ready to do that, you definitely gave us space.
GEORGIO: That was something for me like on the day BA said "guys, I can't believe you don't know how many of these particular mics are in the cupboard. I've worn my boots today. I wore these boots today to kick you all up into gear"… and I went to work going, "ooh, I feel like my dad's just yelled at me"…
ALL: [Laughter]
GEORGIO: … but then I processed it and … took more classes and I'm like, "wow, that kick in the guts has made everyone step up their game".
LENNON: So important.
HAYLEY: Yeah.
ELLIOT: That's a part of the accountability because if he didn't do that, well … he's your boss. If your boss didn't kick you up the butt, he wouldn't be a very good boss.
GROUP: Yeah. Definitely.
LENNON: And I think … as the weeks went on it went from no-one speaking to you literally backing off completely and we just ran it. Like, I think we needed that to get to that point 'cause we weren't there at the beginning.

(Focus Group 1, May 29, 2018)

I then probed as to things that an educator should not do:

LENNON: If you coddled us.
GEORGIO: Too nice, yeah.
HAYLEY: You were in tune with what was happening ... you knew where we were at, what we were learning.... There was that care, you could tell that you wanted us to learn and everything was really intentional, everything was planned out ... you engaged with each of us and where we were at with it and everything like that.
RESEARCHER: Individually or as a class?
HAYLEY: Both.
RESEARCHER: As an educator or as a professional?
HAYLEY AND ALL: Both.

(EdD Focus Group 1, May 29, 2018)

This student discussion highlighted some positive attributes. The educator's (my) engagement in this particular tutorial series needed to be both intuitive and flexible to the situation and the individuals involved. There was doubtlessly a danger if my role became too dominant (doing to much of the work), but it would have been a mistake to overly nurture students since they respected the times I confronted them with the realities of their actions. Sometimes I needed to engage practically: plug in equipment, help with technical problems, work on sounds, parts, instrumentation or perform. This minimised the manifestation of a hierarchal educator/student dynamic and promoted the development of a music production culture. If I became a part of the musical team, this broke down social barriers and promoted the development of profession-based relationships between myself and the students. Sometimes I halted the process and shifted the tutorial into a feedback session where practice could be reflected upon, and sometimes I stepped back and let the students lead and govern the process completely. All of these contexts needed to be actioned instinctively and in reaction to the flow of the pedagogical engagement.

THE EDUCATOR'S LEXICON AND USING HUMOUR

The educator should use a lexicon that students relate to as popular musicians rather than a pedagogical language. Overindulging in technological discussions has a similarly negative impact on PMPP because inclusive PMP classes should prioritise critical thinking regarding context, aesthetic intentions and the socio-musical environment. If the educator can spend some time to get to know the students and develop a lexicon with which the students identify, or else identify as being the lexicon of someone who is a professional, trust can be established, and professional relationships developed. This lexicon must consider the specific cultural context of the student body. For example, in an Australian PMP tutorial, the educator can use phrases like "let's find the groove and sit in the pocket", "can someone please program up a nice phat beat", "hey let's crank it and get some vibe flowing". Using a lexicon such as this can position the educator within a status of musician or industry professional, and this is useful in pedagogical approaches because it locates classroom activities in profession-based contexts.

It is advantageous, then, if the educator is an experienced professional practitioner since the intuitive use of an appropriate PMP lexicon is something that is learned via immersion in many music-making contexts with various types of people and social landscapes. In my classes, we all used sounds as words when describing musical feels and class discussions were centred on current industry-related topics with words that were being used in the profession like "the vibe" and "the tracker process". Eventually some swear words were used as positive affirmations once it was certain all people in the class were comfortable with this. Humour was also used to transform the tutorials into music production cultures. Humour helps to rectify any stagnant educational atmospheres and balances out pedagogical social standings so the educator is seen as a peer or fellow musician as exemplified in Video 17 P (See Video 17 on the website at www.routledge.com/9780367517632).

Alan Williams agrees:

BRENDAN: … and the language that you use in those environments, is it conducive to popular music as opposed to education and academia? I mean, humour was something that we use in the studio all the time, is that something we can use in pedagogical landscapes?

ALAN: Of course.

BRENDAN: Swearing?

ALAN: Yeah … I am somebody who does not tend to use the more formal terminologies and I think in some ways it is from the experience of being an engineer in the studio where you're translating all the time. So I tend to find myself using the language of the streets, of the people.

BRENDAN: Yeah … conducive to the context too, right? Like, I mean you can work on a Hip Hop record, but I've also worked on a lot of Christian records, and I've used different languages.

ALAN: Absolutely. And I think to some degree that's a skill we have to impart to the students. Know your audience, know your environment, know who's hired you and what their response is going to be…. And adapt to that.

(Alan Williams interview, April-May, 2021)

I often use sporting analogies within my lexicon because I find it often helps to contextualise important ideas. Prince Charles Alexander utilises similar techniques:

> I use lots of sports analogies. I'm like take a look at LeBron James. He puts the ball up, it goes in, it looks so easy, but that's thousands of hours of practice and how can you make production look easy without your thousands of hours of practice? Now practice! And that's what we get into in the production and engineering curriculum.
>
> (Prince Charles Alexander interview, April-May, 2021)

Some students may not be interested in sports so aesthetic conversations relating to metaphors can be utilised. Regardless, it is imperative that the educator uses a lexicon that is inclusive for all students, and this will be an intuitive process with every class iteration.

EDUCATOR'S WORKING TOGETHER AND STAYING UP TO DATE

Very rarely will you meet a professional practitioner or educator that is a master of everything. Subsequently, you cannot undervalue the importance of collegiality amongst staff who are working together. Not only does this mean, for example, that songwriting, music industry, performance and production courses can work together with integrated efforts and interests (as is the case at my institution). But educators, by working together, learn from each other.

> So actually, teaching has made me a better practitioner.... Oh, isn't that a great thing to admit? I think the good people teach together, and they certainly embrace ... the idea that I can learn from you, and the student, as much as you can learn from me, and this is an environment where we trust each other. I think there are some educators that don't like that.
>
> (Gary Bromham interview, April-May, 2021)

Gary and I discussed reasoning behind non-collegial approaches and we contemplated a lack of industry experience, ego, fear and power struggles within institutions. Power struggles are terrible things to experience and sometimes bring with them matters of bullying and harassment. Whether you are talked down to, socially excluded, forced to address false accusations or are aware people are talking about you behind your back or in private social media chats, I would suggest the best remedy for such instances are frank, respectful and honest discussions. Honesty, in my opinion, will set you free, regardless of the outcome.

I want to refer to Gary's comment above: "I can learn from you, and the student". During my interview for my full-time position, I was asked by an academic: "how do you intend to stay in touch with all the changes in the music industry?" My answer was:

> that's easy, apart from my own research, all I need to do is establish professional relationships with students where they feel comfortable talking to me about the latest music they are listening too and the students do all the legwork necessary for me to get a good idea of that which is happening in the industry.

The students, and the incredible diversity that comes with an integrated cohort (both sides of the glass), is a tremendous repository of musical knowledge that can be accessed and utilised at will to educate all in the learning environment.

EDUCATOR DEVELOPMENT (RESEARCH)

To conclude this chapter, I want to discuss the important concept of educator development and how this can impact work/life balance. I am one of the very lucky academics who has been granted time by their institution to work on research-related topics. I know this is not the case with the majority of PMP educators, and that many are requested to do so much teaching and administrative duties that research becomes a process undertaken on their own time. University educators all seem extremely time poor and overworked; this was prolific amongst the educators I interviewed, and it brings with it connotations that can negatively influence staff retention.

> So, if I can't develop myself, then I'm not motivated to develop other people, do you know what I mean? And so, there's a huge responsibility on universities, and their business models, [because] obviously the first thing that goes [for an educator] is the research time and the professional development and I think that's a death nail in the coffin if you stop developing your staff.
>
> (Rob Toulson interview, April-May, 2021)

I couldn't agree more with Rob, institutions must value their educator's health and wellbeing, not only does self-development increase domain knowledge, but it gives time for the pedagogical batteries to be re-charged. I have been doing practice-based tutorials since 1998, and the toll these have on my time and energy levels is extensive. In my experience, universities do not allocate sufficient workload time that is representative of the impact these tutorials have on the educator. Like recording sessions, practice-based tutorials require extensive preparation and I go home after a day of tutorials exhausted. Most recently, institutions seem to be requesting the academic progress of their educators, excessively scrutinising teaching workloads and placing more and more pressure on educators to do their own administration, and this negatively impacts staff development.

> I used to go in as a practitioner and the idea was that I brought my knowledge and my experiences. Once it became more academic for me ... the PhD was kind of enforced really in a way and probably the [AdvanceHE] FHEA[1] is enforced as well now. All of these different letters you have to have behind your name...and they want you to be good at admin. [What] I really wanna do post PhD is research-informed practice because I think there's a lot of focus on practice-based research and one of the problems I have with that is, you have to have the hours and hours of practical experience. [I read a lot of books] and there are some that are so flowered up with academic language and I get to the end of the chapter and I think: "I don't even know what this is saying!".
>
> (Gary Bromham interview, April-May, 2021)

I wholeheartedly second what Gary is suggesting. I have reviewed articles and book chapters and have had an identical experience: often chapters are filled with so much verbose language that I have no clue what the person is trying to say. Often this is because they don't either. PMP is a practice-based activity and institutions should ensure that their educators are given the opportunity to engage with the practice of production in an informed manner in order for them to develop an explicit understanding of their practice.

> I think the number one thing to do is to get universities to allow research credit for creative activity for people in production.... I've managed to do that here and as far as I know, I'm like the first person in North America where I no longer have to give print publication.... I do practical things.... When I hand in my CV for performance evaluation, it says like mastering, mixing, producing tracks.
>
> (Jay Hodgson interview, April-May, 2021)

If educators are not given the opportunity to continue their practice, institutions run the risk of losing some very good employees because many value their engagement in their professional practice. As Russ Hepworth Sawyer explained:

> I remember one kid [asked what music work we had been working on]. A co-lecturer said this week "I'm recording this" and another co-lecturer, who was ex-Abbey Road, has been mastering this that and the other. This student said to me, "so Russ what have you done this week?" And I said "well I've been running your course mate, that's what I've been doing." I suddenly realised in that moment that I'd given up something extremely valuable, I'd given up my professional credentials for something that is a permanent job and despite how good the permanent job looks on paper to people, it meant nothing in terms of my kudos to the students. It didn't matter what I said, these other guys were more important.... I left within a year and returned to mastering professionally.
>
> (Russ Hepworth-Sawyer interview, April-May, 2021)

REFERENCES

Anthony, B. (2019). *Perspectives on learning popular music production in higher education from "both sides of the glass"* (Doctorate of Education Thesis, Griffith University).

Cremata, R. (2017). Facilitation in popular music education. *Journal of Popular Music Education, 1*(1), 63–82. doi:10.1386/jpme.1.1.63_1

Frith, S., & Zagorski-Thomas, S. (Eds). (2012). *The art of record production: An introductory reader for a new academic field.* Burlington, VT: Ashgate.

iZotope, (2018). Vocal production, Part 2: How does intention make or break a vocal? (video). Available: https://www.youtube.com/watch?v=RdOHKJNPiAY

Shadyac, T. (1998). (Director). *Patch Adams* (Staring Robin Williams), (Film), U.S.A: Universal Pictures.

NOTE

1 AdvanceHE fellowship are international accreditations for university educators see: https://www.advance-he.ac.uk/

14

Practice-based pedagogical applications

This chapter will briefly outline some useful applications within practice-based PMP tutorials as well as outlining their function and purpose. Imagine these applications occurring within tutorials that are music-making sessions where students are doing the work and the educator is a facilitator rather than exemplifier of the practice. The list will be short and concise but will be flexible enough so that individual variations may be developed where needed within diverse programs and courses around the world.

MONOLOGUES

Tutorials benefit when they are introduced by the educator via a monologue. This approach sets the scene, describes what the tutorial is covering, what the learning outcomes are and what students are expected to do. An example of a monologue can be found in Video 18 (see Video 18 on the website at www.routledge.com/9780367517632). This monologue introduced two provisos: first, the music used for the tutorial would be original, and second, the musical output generated by the class was going to be released by the artist. This forced the students to consider their accountability and positioned them within an understanding of their musical/learning environment.

Monologues are also useful prior to students engaging with more advanced cognitive applications like the producer's vision and producer archetypes because they can be used as in-depth descriptions of professional practice, and this sets important benchmarks for students. However, it is important to keep monologues short; long winded verbalisations and tangential discussions can be confusing and damaging to student learning. The work of Biggs and Tang (2007) suggests that students' attention to low level spoken activity and learning drops off rapidly after 10–15 minutes of engagement. As such, it is recommended that monologues are a maximum of 10 minutes in length.

DOI: 10.4324/9781003055099-18

PRACTICAL ENGAGEMENT AND STUDENT-LED ACTIVITIES

One of the main findings of my research was that, as an educator, my engagement was often too dominant, and I needed to leave a softer footprint on the educational experience. I think it is important for all educators to realise where they can improve. Most of us want to help students learn, but sometimes when we are confronted with passive students, for example, some of us (myself included) have instincts that like to take control and lead. The irony is that learning, with regard to the more advanced concepts of PMP, comes when the students do the work and then reflect together on the experience.

> It is important to establish a culture of sharing knowledge and also having a lot of practical work! The best for me has been when we go through some aspect of production, make something and then listen to it and discuss.
>
> (International student survey response, May, 2021)

Students benefit from practice-based engagements of PMP applications in (facilitated) collaborative learning environments where the music and learning itself can be discussed within the group.

These frameworks should promote student-led pedagogical approaches that reflect peer learning (Green, 2006), "signature pedagogies" (Shulman, 2005) "situated learning" (Lave and Wenger, 1991) and "student as master" (Lebler, 2007). I have defined these pedagogical approaches in Chapter 7, but essentially, they can all be integrated to promote classes where students are in control of the tutorial experience, collaborating within practices that replicate the profession and learning encapsulates the development of an understanding of PMP practice that is inclusive of student reflexivity. However, these pedagogies will not just happen immediately; there is much the educator must do to ensure their development over the course of the students' university experience. The educator is responsible for instilling the necessary confidence in the students via an intuitive, adaptive, and individualised approach that promotes the students' belief in their own abilities. With this set, the educator's engagement (or disappearing act) can be used so student-driven frameworks can develop automatically. Lebler (2007) supports this by suggesting that students

> will need guidance from teachers who have a very important role but one that is quite different from the traditional role of performance expert and mentor. Teachers will need to … provide students with evidence of the benefits for them in engaging with the work of the master while still a student.
>
> (p. 207)

"Class mode" and feedback sessions

When compiling some of my doctoral video materials, it became evident how, as the educator, I instinctively switched to what I call *class mode* from *practice-based mode*. These moments allowed me to step in and discuss important information concerning learning. Following this, students re-engaged in their activities but with knowledge of learning clearly established. This process can be seen on Video 19 (see Video 19 on the website at www.routledge.com/9780367517632). My students reflected on the use of class mode favourably:

> It's good when you [say] 'okay, back to class [mode] for like a second' and then you like switch it off, like, intentionally doing that also really helps 'cause then it's like this is a class when it needs to be.
>
> (Hayley, EdD Interview 3, July 9–10, 2018)

Even though students value pedagogical environments that bridge and simulate the professional space, they also feel comfortable within a class environment because this resembles a safe place where mistakes can be made, and the use of trial and error can be used to explore their creative process.

Tutorials should also include feedback sessions where the group can sit down and discuss how the class related to professional practice, and how the students learnt from this. These feedback sessions are different to the communication that occurs whilst giving feedback and critiquing someone's music. I played a video of a feedback session to my students (see Video 23 on the website at www.routledge.com/9780367517632). In this feedback session my engagement was quite strong because it was early in the year, but I probed students for their opinions on the relevance of feedback sessions in a tutorial:

> Georgio: Well that particular one was really chilled … Like, it's talking about a stressful time but it's chilled expressing it.… You know, everyone's in there pretty open about it … they're just sort of listening.… Also, I think over the year, more people chimed in consistently …
>
> Lennon: When we've had time to go away and kind of think about it and process what we did on that day … a lot more people were vocal about it 'cause they didn't feel it was like on the spot.… I find that for me personally, it's like I just take longer to process things to figure out what I've actually learned … 'cause that … that's just how … I operate as a person.
>
> (EdD Focus Group 2: October 29, 2018, SR)

Feedback sessions during, and at the end of, class allow students to engage, experience and then reflect, and process as a group (see Chapter 10). These discussions can be useful learning situations because students' experiences can be understood and contextualised in terms of those of their peers. If tensions are too high in the session (as was the case in many sessions of my doctoral studies), coming back to discuss things a week later may benefit student learning (EdD Focus Group 1 & 2, May 29 and October 29, 2018).

EXEMPLIFYING COMMUNICATION PRACTICES

The way in which producers communicate with the people in a session is paramount in achieving high-quality results. Subsequently, educators can exemplify this practice so students can learn and apply this craft as their skill in this area improves over time. I discussed this with Lachlan Goold:

> The thing that troubles me the most in it is the timeframes, because … in the studio, how you communicate with an artist is so important … [But you and I] like we're used to giving feedback…. You can't just unload on someone when they're singing…. If you unload all [the things they are doing wrong] on someone, the next time they go to sing, their head … goes so far away from what the song is about and why they're actually there in the first place that they're never gonna give a good performance…. So, to teach those sort of approaches to students, you can only ever pepper it in, like really give them little bits of it and hope that they run with it [in their practice].
>
> (Lachlan Goold interview, April-May, 2021)

By exemplifying communication practices and then encouraging students via communication mentorships, education in this area can be peppered in regularly, as Lachlan suggests, over a sustained timeframe. This approach is necessary because these are tacit soft skills that we must nurture in students, and these may take time to develop. An example of my mentorship of a student in communication of this type is exemplified in Video 20. This particular student wanted to work as an engineer/producer and asked for assistance with his communication skills. Over time I developed these with him during class tutorials, as can be seen in Video 20 (see Video 20 on the website at www.routledge.com/9780367517632).

Toby Seay engages in this practice also:

> Toby: I say to my students … "watch this, I'm about to push the talk-back button …"
>
> … You know, and so I'll push it and I'll talk to the musician. Then I'll turn around. It's like [I ask the students] "what was I actually saying to them?" you know? [I like to] have those discussions all the time. And you know, we'll [learn to] be nice, [we all might hear something and say] "wow man, they're just really not in time". [And I will ask students] "Okay, what can we tell them" [to fix that]…. And so, that [type of] communication I try to stress a lot …
>
> (Toby Seay interview, April-May, 2021)

TURBOCHARGED LEARNING

My research has developed a new pedagogical approach labelled *turbo-charged learning*. I have found, with students in the advanced years of study, that it was necessary for the educator to *kick-start* processes. This occurred usually when the students were struggling to make a decision or the session was losing focus. This kick-start functions like a burst of energy that helped the session progress further. Upon reflection of this activity whilst driving home after teaching one day in 2018, I felt the turbo of my diesel-powered four-wheel drive kick in to give my ute the added boost it needed to accelerate. This analogy stuck in my head. The educator's role, in these final advanced classes, was like the turbocharger of an engine. When the engine (the tutorial and students) needed an added boost, the educator (the turbo) injected the necessary professional advice to get the tutorial moving again. Rob Toulson suggested he does the same thing: "yeah, that kind of thing where you just stand and observe and obviously join in, and give some production advice if there's something that needs some help" (Rob Toulson interview, April–May, 2021). An example of the process in action is exemplified in Video 21 P (see Video 21 on the website at www.routledge.com/9780367517632).

Turbocharged learning is what the educator's role can develop into over time. Classes can shift from formal and pedagogically oriented to more professionally framed through a kick-starter when needed. This analogy (turbocharged learning) is reminiscent of some discussions on facilitation I provided previously where educators respond to students' needs by giving and removing assistance. These are very different to class mode interjections (focussed on learning). Rather, turbocharged interjections are practice-based suggestions that are designed to kick-start (reorient) the music-making process back towards a profession-based context. Again, this reinforces that the educator needs to be an experienced music producer so as to have the knowledge and confidence to suggest suitable musical, technological or creative practice options. Students often affirmed that this turbocharged approach exemplified my practice towards the end of tutorials, as seen in Video 22 P (see Video 22 on the website at www.routledge.com/9780367517632) and discussed in focus group activities:

RESEARCHER: So … puttin' you under the pump [when I went], "go and do that, you guys". It helped?

LENNON: Yeah. Well I know that if … if you hadn't pushed us to do it, I would've been banging my head against a wall for hours at home …

RESEARCHER: So, what do you think would've made you guys go, "you know what? We need to go in the other room and do that" rather than me saying, "hey guys, go in there and do this"?

HAYLEY: Or it would've taken us longer. I feel like there would've been a lot of beating around the bush, like, "what should we do? …"

ELLIOT: Yeah … I do think having those pushes and … in … almost like showing us different ways that production sessions can go, 'cause we … didn't really know that this was a possibility … if this was the most effective way and it clearly worked quite well.

HAYLEY: Um … is it the, what is it turbo …?
RESEARCHER: Turbocharged learning.
HAYLEY: Turbocharged learning. I think … you had prepared us enough throughout the past few weeks and months … with the tools and confidence … but sometimes we did just need that extra push …
RESEARCHER: You needed that turbo to kick in …
HAYLEY: Yeah, and then we're like, "oh yeah, of course, like, that makes sense, we'll go and do that".

(EdD Focus Group 2: October 29, 2018)

"ACTION LEARNING", ETHNOMETHODOLOGY AS A TEACHING AID

To conclude this chapter, I want to highlight that the educator should intuitively adapt practice-based tutorials in reaction to student engagement. As such, "action learning" (Biggs & Tang, 2007) is a useful pedagogical tool. As described in Chapter 7, "action learning" allows the pedagogical delivery to adjust systematically in reaction to classroom events. By working with students as a musical peer, the educator, in this scenario, has a vantage point where they can witness and document improvement in student practice. During pedagogical applications, I also use research informed practice to adjust delivery; ethnography has also been useful to document personalised learning experiences of students and adapt pedagogy on the fly. Barton suggests that participants often do not recognise the complexity of the change they go through, and it is the job of the ethnographer to make a strong connection with the people with whom they work. "It is a reflexive relationship or at least it should be. In a way the role of observer slowly morphs into participant and then initiation into being an accepted 'member' of the social/cultural phenomena being investigated" (2014, p. 15).

The PMP educator can use "action learning" and ethnographic processes to inform their pedagogical engagement by becoming a participant observer of the music being made and the learning that is occurring. Throughout my doctoral study I kept a researcher's journal that documented individual and group student progress. I found it most useful to read the journal and adapt classes in reaction to these notes. This is an approach I have maintained since 2018 and I advocate its utility in promoting flexible learning frameworks.

REFERENCES

Barton, G. (2014). Ethnography and music education. In K. A. Hartwig (Ed.), *Research methodologies in music education*. Newcastle upon Tyne: Cambridge Scholars Publishing.

Biggs, J. B., & Tang, C. S. (2007). *Teaching for quality learning at university: What the student does* (Vol. 3). Maidenhead: McGraw-Hill/Society for Research into Higher Education & Open University Press.

Green, L. (2006). Popular music education in and for itself, and for 'other'music: Current research in the classroom. *International Journal of Music Education, 24*(2), 101–118. DOI: 10.1177/0255761406065471

Lave, J., & Wenger, E. (1991). *Situated learning: Legitimate peripheral participation.* Cambridge, MA: Cambridge University Press.
Lebler, D. (2007). Student-as-master? Reflections on a learning innovation in popular music pedagogy. *International Journal of Music Education, 25*(3), 205–221. DOI: 10.1177/0255761407083575
Shulman, L. S. (2005). Signature pedagogies in the professions. *Daedalus, 134*(3), 52–59. DOI: 10.1162/0011526054622015

15

Collaboration and creativity in popular music production pedagogy

Much of the popular music education literature I have cited in this book has alluded to the value of collaborative learning frameworks (see Chapter 7). I have also promoted collaboration in many of the chapters prior (see Chapters 9–11 and 14) and cited literature from a broad scope of areas. In this chapter, I will discuss the ways collaboration can facilitate learning in a multitude of ways. When undertaking PMPP, simply arranging students into groups is not going to result in magical outcomes. In contrast, collaboration must be planned and designed for diverse student learning and it is important to understand the different ways in which collaborative engagements can add value to PMPP.

THE BENEFITS OF COLLABORATIVE LEARNING: ENGAGEMENT, A COLLABORATIVE HUB, PERSONALISED LEARNING AND A SENSE OF COMMUNITY

Music making is rarely a solo activity. As such, it is essential that PMP education must include collaborative experiences so musicians get used to working with a broad range of people. One of the key targets in collaborative frameworks is the development of social skills and educators must find innovative ways to develop a student's tacit abilities in these areas. Donohue (2021) discusses the utility of education that engages vocalists in collaborative frameworks with both audio technology and "the human beings who work in that medium" (p. 137). Donohue goes on to describe a three-step pedagogical process that includes: (1) education regarding technology and the role of those that use it, (2) hands on experience (for vocalists) in working with technology themselves and (3) communication strategies between vocalists and audio engineers that promote knowledge transference and a spirit of gratitude (see pp. 139–142). This work resonates strongly with my earlier sentiments regarding the value of integrated cohorts where music technology and popular music students work and learn together. Music production cultures can be used in these scenarios to nuance collaborative learning from varying perspectives: working with Hip

DOI: 10.4324/9781003055099-19

Hop or, in contrast, a Motown inspired track (see Anthony, 2021). This not only helps students to learn how to communicate in different music-making scenarios but also to work with music that may be outside of their comfort zone where they will need to rely on the strengths of others. In these instances, learning occurs in many ways.

My research has shown that students found value in collaborative frameworks because learning tacitly occurred alongside student engagement (doing). However, it also occurred by watching, talking and listening to others.

> I know that, for me, I pick up the most … when I am actually doing something, when I am practically involved.
>
> (Lennon, EdD interview 1, February 26–27, 2018)

> I think, also, because there was a lot of practicality, like actual working through stuff, that helped as well. Working together as well … I think because you have a bunch of people who are the same as you learning together, it just strengthens that like learning experience.
>
> (Hayley, EdD interview 1, February 26–27, 2018)

I played a video of a collaborative tutorial to my students (see Video 6 on the website at www.routledge.com/9780367517632, 4:48–9:52) and I asked them to discuss the importance of this collaboration:

HAYLEY: It's the little details and tweaks where you just need somebody else's input sometimes …

LENNON: That's why Hayley needed to take control of that session because I was stumped with it … [what] we came up with together was just so much better than what I had.

HAYLEY: And I wouldn't've … been able to have written from scratch. It was cool to take your ideas as well, so it was a really cool process for everyone.

(EdD Focus Group 2: October 29, 2018)

The students' comments above reiterated that there is strength in collaborative learning models, especially when students engaged with practices that concerned creativity. Similar to Christensen's (2017) discussions on the usefulness of co-creation activities with rock musicians, the concept of a *collaborative hub* seems to have given my students the foundation to engage in creative practices because a collective effort draws on greater domain influences that can help shape the result (Thompson and McIntyre, 2013). This collaborative space is similar to the alliance that professionals form when working, and it bonded students and produced an environment in which participants suggested they had "become more of a team" (Hayley, EdD interview 2, April 24, 2018). When opportunities for critical reflection and discussion were added to these collaborative engagements (as part of a facilitated learning experience), group understandings were generated, and learning ensued for all.

Collaborative frameworks also benefit when the educator becomes an active participant in the music-making experience – a part of the team. This does much to shift the students' mindset towards one that sees themselves and their educator on even footing and capable of making music in diverse collaborative frameworks. Bryden Stillie elaborates:

> We need to get students to a point where they understand that they can do it for themselves, because they're not always going to have somebody over their shoulder, or in a supervising [capacity saying] "how are you going with that? Where do you think it's going to go next? What sort of support would you need to do that?" They need to understand the questions to ask themselves.
>
> (Bryden Stillie interview, April-May, 2021)

Personalised learning

I also asked students of my doctoral research to reflect on how collaborative frameworks enabled personalised learning. Thom suggested that his interactions with a group helped him identify what he was doing right and what he was doing wrong. Hayley suggested that she learnt a lot in collaborative frameworks by watching others learn and thereby picking up on things that she may not have if she was working individually. Lennon suggested that learning in a group setting allowed her to feel a sense of security in her learning because the students were all in it together. Finally, Georgio highlighted the importance of communication between both sides of the glass and suggested that group contexts illuminated parts of his communication process that needed improvement (Personal correspondence, November 1, 2018). It seems, on an individual level, that by working and collaborating, participants gathered an understanding of how others work, and this helped them reflect on their own positive and negative attributes as music producers.

A sense of community

When I spoke to Jarritt Ahmed Sheel, we discussed collaboration in Hip Hop music production:

> One of the cornerstones of Hip Hop [that is often implied] is the topic of community. It's often an unspoken thing because it's a Hip Hop community, that's what makes it great.... Because nobody can say what the community does, the community on the bottom says what is hot.
>
> (Jarritt Ahmed Sheel interview, April-May, 2021)

Aligning with Jarrit's comments, I would argue that a sense of community underlines many popular music genres (Punk and Progressive Rock, for example) and is essential for high-quality PMPP to occur at HE. However, Jarritt also suggested that many Hip Hop musicians do not undertake HE study because they do not deem it necessary. "They say: 'I know my lane and I don't need that validation because that's not putting money in my pocket'. Or 'that's not my community so I am not doing it for them anyways'" (Jarritt Ahmed Sheel interview, April-May, 2021). In my opinion this is one of the resounding issues HE institutions face: they are often not perceived to be a valid part of the community. It is important to recognise that some music types, like Hip Hop, are developed in opposition to societal norms, like university. This problem is exacerbated further by the fact that Hip Hop is currently one of the most consumed types of music on the planet (Nielson, 2018) and, therefore, HE has a responsibility to offer an enriching experience for Hip Hop musicians, should they wish to attend. This is a complex problem and there is only so much any one institution can do to tailor to the needs of all aspiring students. However, I know that many institutions (including mine) are contemplating ways that their programs can be adapted so that they are deemed as valid to such pop music communities. Cultural change is required as this is a multifaceted problem that requires attention in realms that include diversity and inclusiveness of teaching personnel, curriculum, entry requirements and overarching program and course focus (away from placing hierarchy on Western classical music, for example).

LEARNING JOURNEYS IN COLLABORATIVE FRAMEWORKS

Many of my past students and some educators that I have spoken with find interdisciplinary group collaborative frameworks challenging. These can be tough social contexts to traverse, and often students find it difficult to produce high-quality musical outputs when working with different types of people. However, I would argue that these socio-musical interactions are important for students to experience so they may diversify their music-making capabilities and develop their soft skills. Collaborative workflows are versatile if they are also viewed as *learning journeys*. Learning journeys are music-making activities where students are tasked with engagement, critical enquiry and adaptation to emphasise the completion of a musical project rather than judging the quality of the musical work. The integral element here is that the student needs to apply critical thinking and reflection as a by-product of their activities and to promote problem-solving initiatives. Problem-solving is a transferable life skill that PMP students will need post-graduation and learning journeys provide a wonderful resource to develop this. Bryden Stillie agreed:

> I think that building the collective responsibility to get projects completed first and foremost, which is kind of, if you're working in an ensemble or any kind of professional capacity with music, often what you've got to think about, is well, 'I'm part of a bigger system that's happening here'.
>
> (Bryden Stillie interview, April-May, 2021)

PMP pedagogical approaches need to consider new ways to engage collaboration so that the onus is taken off the educator to facilitate learning. This helps to remove any reliance on the educator that will not be present after graduation and also any musical bias the educator may passively impart through their subjective preferences. Learning journeys accomplish this effectively because they can be undertaken primarily within autonomous student applications. This places the student in control of the music-making activity and encourages them to engage with project management activities as well as affording them the musical freedom to work with musical styles that they find engaging. This allows pedagogical approaches to cover musical styles that are difficult to teach in group classroom settings (Hip Hop, for example). Students can organise their autonomous engagement; be involved in creative discussions with their peers; work their way through the communication, writing and production process; and, if needed, have regular check-ins with their class to obtain feedback on their work. An example of this type of pedagogical approach that promotes electronic and online collaborative frameworks can be seen in some of my more recent scholarly publications (see Anthony, 2020).

THE NEW AGE OF COLLABORATION

In a questionnaire I sent Nyssim Lefford, she brought up some interesting points regarding the concept of collaboration and why we must consider this through multiple lenses (the educator's and the student's).

> Technology has, in essence, enabled a kind of self-actualization.... Production work (in the eyes of many students) does not (now) require a diversity of experts who can contribute the necessary components. Many teachers of PMP have a very different view of collaboration. Most teachers learned to work with others, to collaborate, because the work most often demanded the contributions of [many]. Also, for the teachers, the technology needed to produce was largely not available outside of professional environments anyway. So, the teachers had no choice but to learn how to play a part in a bigger operation merely to access the tools.
>
> (Nyssim Lefford questionnaire response, April-May, 2021)

Nyssim went on to explain that the large recording studio music-making environments that many educators experienced prioritised the quality of the work produced and the delegation of tasks as the primary concern. Quality was the service being sold by large facilities and the subjective experience of the individual was important but often secondary. However, and in contrast to this, for many students, music production can occur in small collaborative scenarios (1–3 people) and high-quality audio can be achieved in an assortment of studio environments (including their home). Students value work if it feels fun and meaningful; their subjective experience is often of primary concern. These differences between educators and students can often facilitate confusion regarding collaboration because teachers often expect students to appreciate the wisdom of delegating tasks in large group contexts. These approaches are not always obvious to students because they can be preoccupied with how they feel and their activities within the social experience of collaboration. Students naturally prioritise the emotional aspects of music making, they aspire towards activities where it feels they are working well with people. Certainly, these aspirations have merit in modern music making; it is important that professionals get on with clients, collaborators and colleagues (Nyssim Lefford questionnaire response, April-May, 2021).

> One has to consider what others might be thinking/feeling, and this does not come easily to all students. Furthermore, clients want both a good working experience and a good work product. So, one has to consider the social situation while also considering the goals of the work and what is necessary to achieve the desired results. This requires complex thinking, which also does not come easily to all students. In many student collaborations, the investment in the social experience is high. They avoid difficult discussions and compromises so as to not risk sociability ("likes"), comfort and group happiness, but this is frequently at the cost of the product. Teachers may be inadvertently encouraging this because it can be quite difficult to guide students' thinking about trade-offs, compromises and managing difficult discussions with collaborators.
>
> Many social and technical developments have also led students to be suspicious of traditional notions of authority in a production environment. They tend to reject the idea that hierarchical arrangements could ever be desirable. It is not at all a bad thing to reconsider power structures in PMP given the long history of abuses.... However, this also may be leading to situations where there is little regard for the value of others' experience and accumulated expertise. It can create a situation where 20+ years of experience is no more valuable than 10 weeks – perhaps because observing those differences could be uncomfortable.
>
> (Nyssim Lefford questionnaire response, April-May, 2021)

If, as Nyssim suggests, students and educators have different perspectives on the value of collaboration (social interaction, performance, quality or the bringing together of experts), this only justifies the necessity for deep contextualising discussions regarding this in the classroom. I also want to highlight Nyssim's words regarding the student's avoidance of creative conflicts:

> One has to consider what others might be thinking/feeling, and this does not come easily to all students.... [O]ne has to consider the social situation while also considering the goals of the work and what is necessary to achieve the desired results. This requires complex thinking, which also does not come easily to all students.
>
> (Nyssim Lefford questionnaire response, April-May, 2021)

I would attest that these collaborative engagements (that target soft skills) are some of the most important work students can do in preparation for their futures after HE. However, students will not simply traverse these creative conflicts at first attempt, rather they will require guidance on how to communicate, how to read people and how to use psychological approaches to get the project to completion. Nyssim and I discussed collaboration at length and we both felt it necessary that students are given instruction on understanding varying perspectives of collaboration (modern, historical), why it is useful for them to engage in different methods and how collaborative frameworks can develop trusting relationships that generate high-quality material.

> I mean [collaboration can be] having somebody who's willing and able to inspire you, right? There's a place for being inspired and feeling safe and being happy and being really into [the experience of the work], right? And hopefully ... we spend a lot of time in that space. But there's also room for being thrown out of your comfort zone [for the purpose of exploring new territory]. And, when you're thrown out of your comfort zone, you really have to trust your collaborators. Trust is something that students need to learn how to earn.
>
> (Nyssim Lefford interview, April-May, 2021)

Collaboration in PMPP therefore, not only provides the opportunity for peer learning, learning journeys and social skill development but it also aids the development of a graduate's perception of collaborative possibilities that will be valuable into the future.

FACILITATING AN EDUCATION IN PMP CREATIVITY

First, I have already discussed PMP creativity literature in Chapter 5, so I refer back to this at this point. Second, I have positioned this section of the book in this location because I see it sitting best within discussions of collaboration and practice-based applications. This is primarily because, in my experience, students learn about creativity more readily if it is approached in a group context. Learning about PMP creativity in groups also takes the onus off romanticised notions of an individual's musical intellect. Earlier, I alluded to some data that demonstrated how very few graduates of HE PMP courses felt they had developed any solid understanding of creativity. Paul Thompson and I discussed this problem:

> Part of the problem that we have sometimes in popular music is that we have educators who aren't popular musicians, they are classically trained musicians who like a bit of popular music. Or they're jazz musicians who think that popular music is the easy version of jazz.
>
> (Paul Thompson interview, April-May, 2021)

The problem Paul alludes to above is that some PMP educators have a limited understanding of popular music and PMP creativity. However, my research has shown that students find the courage to engage with their subjectivity gap and the creative process more actively once they have a basic understanding of creativity. Aspiring music producers hunger for knowledge that unpacks the creative process. As Paul Thompson suggests:

> We have music analysts, in popular music.... But ... that's not where the cool bit happens in popular music. The cool bit happens in and around why people do it and how they do it. Brendan, if we didn't care, we wouldn't bother watching mix with the masters, or we wouldn't bother going to conferences and go, oh how do you do it? We are fascinated by, you know, any good butcher, any ice cream maker, any welder, plumber, will watch other people do what they do and go, oh that's a good idea, that's cool.
>
> (Paul Thompson interview, April-May, 2021)

Students are just as keen to learn these things, and I have previously tabled Paul's and Philip McIntyre's work on PMP creativity (see Thompson and McIntyre, 2013) as valuable literature in this area. I discussed with Paul how he utilises the systems model of creativity (see Figure 5.1) in the classroom:

> I've redesigned one of the modules, which is called creative music production, around the creative system. So it holds the mirror up to them [the students] to say, "okay what is it that I do as a music producer" ... We have singer/songwriters in the course, we have electronic musicians, so you can apply that framework and use it for them to explore their own creative practice.
>
> (Paul Thompson interview, April-May, 2021)

Paul and I went on to discuss how students arrive at HE with an enormous amount of knowledge, but they don't realise it. Instead, they're waiting for us to tell them about something that *they* specialise in:

> They know way more than me, on a subconscious level. So, part of our remit, I think, is to remind them that they have a shedload of fantastic knowledge. The second part is where they struggle: the way to formalise and externalise it, and so I ask students to go through a process of analysis. This is where they break down the stylistic components of the musical style that they're going to make. It basically gives them a framework to hang their coat on and go, "okay, yeah, I know way more than I thought". Then we send them through a process of synthesis, so they take all the little bits they've got and bring them all together in a new way. They create a system, then at the end of it, they evaluate it, and they compare it back to the original analysis. Does it stand up? What does the field say? ...
>
> I made a conscious decision ... to say, "okay we need to serve our students better, we need to lead them by the hand through a process, so they come out enriched at the other end". It feels like a magic trick, all we're doing is showing them what they already knew, it's just that we're allowing them to formalise it and externalise it in a meaningful way.
>
> (Paul Thompson interview, April-May, 2021)

The course Paul is discussing is a postgraduate offering (Master's level) and, accordingly, the systems model is rather complex. To address this, I have revised the systems model into a simplified and student-orientated version for PMP undergraduate courses. My undergrad systems model (as shown in Figure 15.1) integrates Csikszentmihalyi's (2014) systems model of creativity with the student's university experience. In this model, the three systems are represented by the student's aural library (domain), their agency (including vision and the archetypes they aspire towards) and the validation of their music by their peers and the industry (field). This model works well as a fundamental introduction to PMP creativity in undergraduate degrees.

I developed the PMP undergrad systems model from *Kerrigan's revised system* model and the *Systems Centred Learning (SCL)* model taken from McIntyre et al. (2018). Kerrigan's (2013) revision of the original systems model provides a centre section that is given over to *creative practice.* McIntyre et al suggest that "by allowing practice to take centre stage, the model visually de-centres the individual, the domain and the field as the possible loci of creativity" (McIntyre et al., 2018, p. 112). I have done similar in my undergrad model above.

The SCL model of McIntyre et al. (2018) can be used to inform curriculum development where there is (1) a domain of knowledge in this case termed a course, (2) a field of experts called a teacher and peers and (3) a group of agents – students (p. 114). The authors outline how this model of creativity has utility in educational frameworks:

> A course (to use the generic term) is, from this perspective, seen as a subset of a domain of knowledge. The material covered in the course is

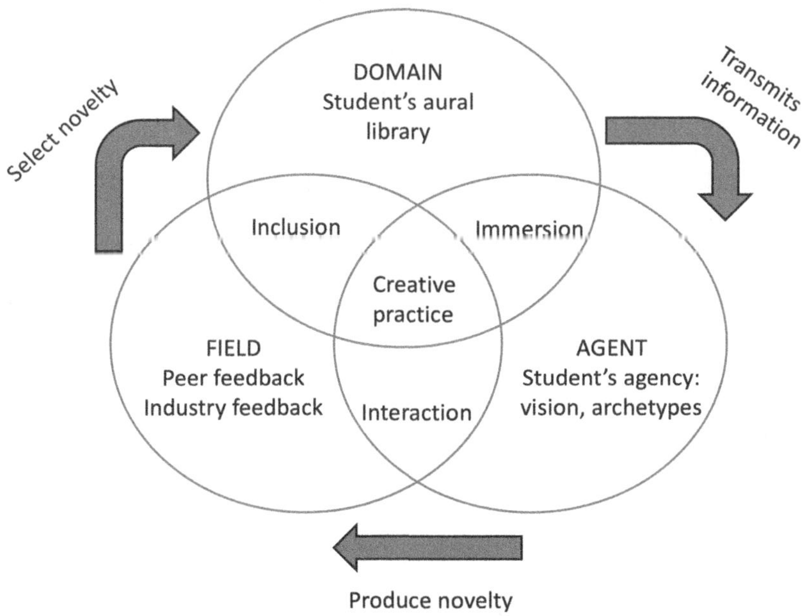

Figure 15.1 Popular music production undergrad systems model of creativity – the student's lens (Adapted from Kerrigan's revised system model and the Systems Centred Learning model taken from McIntyre et al. [2018]).

> selected by the teacher who is a member of the field and they reference the wider community of scholars and industry practitioners who constitute the field. The teacher initially defines tasks, projects and assessment events for the student to study and works with the student on these tasks. The student, seen as an active agent with an individual background and unique set of experiences constituting their habitus, accesses the course content, interacts with the teacher and engages with the wider field and domain of knowledge. The SCL model also provides a rationale for the development of assessment events within a course, some of which focus on domain acquisition whilst others require evidence of engagement with the field. Overtly explaining this approach to assessment design to the students builds an understanding of the relevance and integration of assessment to desired learning outcomes.
>
> (McIntyre et al., 2018, p. 126)

The SCL model outlines a reputable and robust method that may be applied to the design of curriculum and assessment items for many PMP courses, an example of which will be presented in Chapter 16. Uniformity in this regard would help to lay a foundation of practice that we all may utilise as influence.

SUMMARY

The greatest disparity in learning outcomes I have found from students who participated in the international student survey have come in the areas of creativity, emotional transference, communication and psychological skills and electronic music production. In my opinion, these are necessary areas for educators to consider if they are to call their courses "music production" as opposed to "sound recording". Without unpacking the realities of PMP creativity to students, educators run the risk of placing them blindfolded into the world of creative practice. Some may stumble around and happen upon a framework that works even if they have little understanding of how or why it works. More rigorous pedagogical approaches that provide education regarding PMP creativity empower students to develop a greater understanding of practice *while* they practise. There are many ways to go about doing this, I have outlined some above, but feel free to develop your own. My research has shown that an understanding of PMP creativity brings with it students who are more confident and willing to participate and collaborate.

REFERENCES

Anthony, B. (2020). The tracker producer and top-liner collaborative music production project. In A. Bell (Ed.), *The music technology cookbook: Ready-made recipes for the classroom*. (pp. 275–281). New York: Oxford University Press.

Anthony, B. (2021). The impact of diverse music production cultures on popular music pedagogy in higher education. In S. Holley, K. Reinhert & Z. Moir

(Eds.), *Action-based approaches in popular music education* (pp. 107–115). Denver, CO: McLemore Ave Music & Fflat Books.

Christensen, T. N. (2017). Teaching co-creation: Paradoxes in rock and pop ensembles. In T. Chemi & L. Krogh (Eds.), *Co-creation in higher education: Students and educators preparing creatively and collaboratively to the challenge of the future* (pp. 205–223). Rotterdam: Sense.

Csikszentmihalyi, M. (2014). The systems model of creativity and its applications. In D. K. Simonton (Ed.), *Wiley's book of genius* (pp. 533–545). Chichester: John Wiley & Sons.

Donohue, L. (2021). An essential collaboration: Teaching vocalists to work with audio technology and engineers. In S. Holley, K. Reinhert and Z. Moir (Eds.), *Action-based approaches in popular music education* (pp. 137–143). Denver, CO: McLemore Ave Music & Fflat Books.

Kerrigan, S. M. (2013). Accommodating creative documentary practice within a revised systems model of creativity. *Journal of Media Practice, 14*(2), 111–127. https://doi.org/10.1386/jmpr.14.2.111_1.

McIntyre, P., Fulton, J., Paton, E., Kerrigan, S., & Meany, M. (2018). *Educating for creativity within higher education: Integration of research into media practice*. London: Palgrave MacMillan.

Nielson. (2018). 2017 U.S. Music Year-End Report. Retrieved from https://www.nielsen.com/us/en/insights/report/2018/2017-music-us-year-end-report/

Thompson, P., & McIntyre, P. (2013). Rethinking creative practice in record production and studio recording education: Addressing the field. *Journal on the Art of Record Production*, (8). Retrieved from http://arpjournal.com/rethinking-creative-practice-in-record-production-and-studio-recording-education-addressing-the-field/

16

Learning popular music production

In this chapter I will present a diagram that exemplifies the overarching pedagogical approach that I propose can be utilised in diverse HE programs. This educational framework seeks to develop a student's knowledge of the greater practice of PMP and relies on an integrated cohort and the student's prior learning so that every aspect of PMP may be experienced, critiqued and reflected upon.

ACQUIRING DOMAIN KNOWLEDGE AND ENGAGING WITH MUSIC PRODUCTION CULTURES AND PEER EVALUATION

PMP can be learnt through a cyclic framework that is an adaptation of the creative systems models of Kerrigan (2013) and McIntyre's et al. systems centred learning (SCL) (2018). First, students obtain advanced domain knowledge in lectures, learning tutorials and feedback sessions before undertaking practice-based activities in diverse music production cultures that are informed by their program of study (making music). These activities can occur both in class and during students' autonomous work. Second, through critical reflection and feedback with their peers and members of the profession, the student's practice can be evaluated. The evaluation process can take place within assessment and peer reflection sessions and not only validates students' music but also their process and agency as practitioners (including social skills) from envisioning to completion. Through many cyclic iterations and engagements with diverse music production cultures (online, acoustic and electronic), an understanding of the many elements that are encapsulated in the greater practice of PMP is thereby developed. This understanding is nuanced by the student's individual affordances and aspirations as a popular musician. The pedagogical approach is exemplified diagrammatically in Figure 16.1.

The onus of the learning experience revolves around the student's engagement with diverse music production cultures and creative practice facilitated through their autonomous and class-based music-making activities. The three elements constantly influence the music production cultures

DOI: 10.4324/9781003055099-20

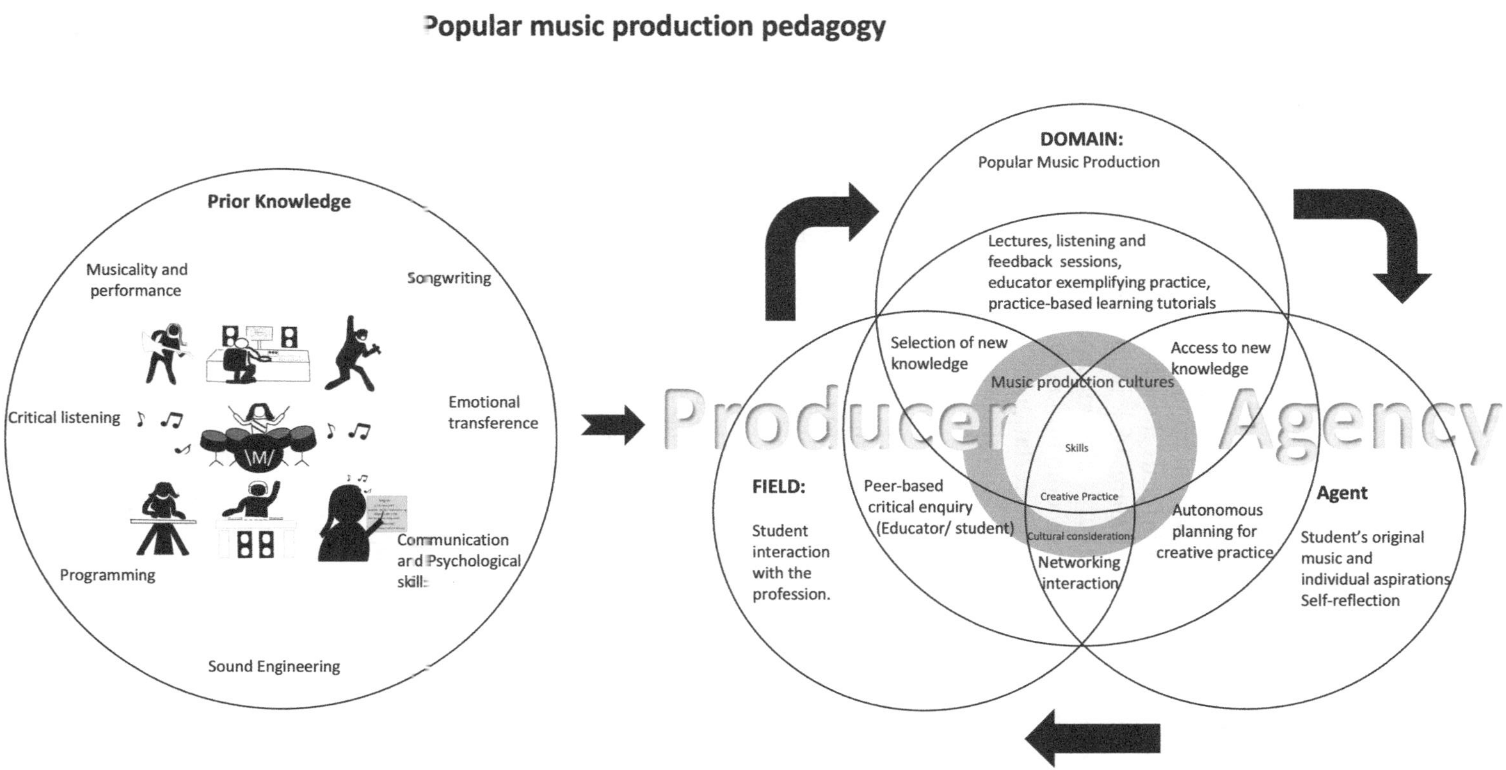

Figure 16.1 Popular music production pedagogy (Adapted from Kerrigan, 2013, p. 114; McIntyre et al., 2018, p. 113).

with which students participate. The domain element provides (weekly) information that the student can access and choose to apply and innovate upon with their own volition. The students' practice and music are regularly validated by peers, the general public or the industry through feedback sessions and public release strategies. Sometimes student work will not be validated, this could be an outcome of low marks in an assessment item, or exemplified in feedback the students receives. Through critical reflection, students can adapt their process, undertake new pathways so as to improve their work and increase their knowledge regarding the making of music. This model is flexible enough to work with both coursework activities and project management applications where an engagement with industry and the consumer informs the validation process.

An interesting note on Figure 16.1 is that producer agency is located within one's engagement, but it also incorporates all three elements. This is because the acquisition of domain knowledge (staying up to date with that which is current), the manifestation of musical outputs and interaction with the peers and public who validate one's work is part of the portfolio musician's practice. Music making does not exist without a constant sense of growth tethered to the acquisition of knowledge, making music and interacting with the field to seek validation. The combination of these factors aids the development of an understanding of PMP as a form of practice. Students don't just learn how to do PMP, they learn how to be PMP practitioners. Student survey responses positively reflect on this process in action:

> Every session I've had since these classes has 110% been positively impacted by the practices of the course. These classes have helped me develop invaluable habits and techniques to apply in everything – including things like engagement, critical listening, leaving your ego at the door, using comp sheets, counting bar numbers, reading people's body language, networking – the list is almost endless.
>
> (Student response, EdD post-survey)

> My external sessions were definitely positively impacted by what I have engaged with in the tutes, especially with the tracker assessment and my practice of top line writing.... I also thought these tutorials transformed my mentality and thinking, to think like a producer instead of just a singer/songwriter.
>
> (Student response, EdD post-survey)

This framework enables learning regarding the expansive practice of PMP, but it is flexible and can be tailored to the individual student on many levels. Below I will provide some isolated examples of how this is being facilitated before discussing the integral role of critical thinking in PMPP.

Learning in action: Communication

RESEARCHER: Any other things that you feel like you're improving at?...
THOM: Communication, I think we can both agree on that.

RESEARCHER: How do you think you've improved?
THOM: ... just being able to understand how they feel through the glass and what they want and just communicating what I'm doing more instead of just not saying anything through talkback...
RESEARCHER: And how are you learning [this]?
THOM: ... Ah, well I've been learning pretty much by watching other people.

(Thom, EdD interview 2, April 24, 2018)

If we relate the above discussion to Figure 16.1 you can see how learning is occurring in this instance. Thom engaged in music production cultures and, then, during peer reflection (field element), it was recognised that he needed to work on his communication (his actions in this area were rejected by his peers). This then became new working knowledge for Thom (domain element). Through his re-engagement in collaborative music production cultures (agent element), Thom watched how others were communicating and learnt from this (domain element). He then actioned this in his practice and through discussions with me and his peers (field element) his improvement was validated (domain element).

Learning in action: Pitch presentations, targeting briefs, discussion and feedback

I often get my students to pitch their ideas for PMP to a peer group (students or industry partners) prior to their engagement with music making. Most of the times I get students to target briefs that I have provided for them (oftentimes based on publishing briefs I received as a songwriter), or that have been provided by industry partners. Not only does this engage with elements of project management where musicians will pitch their product to clients, it also requires the student to critically analyse their process. A student pitch presentation will be informed by the domain element (lectures, listening sessions, past tutorial experiences, a student's aural library, etc.) and normally includes considerations of the student's intentions, reference material, songwriting (the function of the song and its lyrics), creative practice options (the technology being used) and any socio-musical considerations (people involved).

When contextualising the pitch presentations into Figure 16.1, they exist in between the agent element and the field element (networking and interaction) and bypass music production cultures until ideas are recognised by the field as valid (field to domain element). These interactions are reminiscent of creative meetings at advertising agencies or record or publishing labels where musical references and demos are played and individuals promote their ideas to the community to gain approval. Sometimes approval, in some or all aspects, may not be forthcoming, and students may need to critically reflect on possible adjustments. Once this has occurred the necessary music production cultures can be designed and music made in accordance with the pitch session (agent element). The music and the agency of the participants can then be evaluated by the student's peers and the general public (field element). In this style of engagement, students not only learn whilst operating as the producers of projects, they also learn tacitly while being the providers of feedback. In this setting, they process how others go about their creative processes, at all stages and in all roles, and this helps to broaden all students' domain knowledge.

These reflections on learning highlight the usefulness of *active participatory learning* (Ciobanu, 2017). Active participatory learning declares that "the didactic framework must be able to create situations where [students] are forced to use a vast array of processes and mental operations so that they have the opportunity to use the material taught to solve the given tasks" (Ciobanu, 2017, p. 167). I have been using this kind of pedagogical approach for the last 20 years because it resonates with how I learnt music production in the professional world. It is best summed up by suggesting that the student is encouraged to use their

> ...intrinsic curiosity, the desire to observe, explain, explore, discover. Active and participatory are those methods that mobilize the student's energies that help him concentrate his attention, stimulate his curiosity. Active-participatory methods emphasize operational knowledge, learning by action; bring students into direct contact with real life situations.
>
> (Ciobanu, 2017, p. 167)

CRITICAL THINKING: LEARNING HOW TO CONSIDER MUSIC MAKING IN DIVERSE WAYS

To conclude this chapter, I want to focus on the integral issue of critical thinking. A majority of the educators I interviewed for this book supported the notion that in some way "a university education should create critical thinkers" (Nyssim Lefford, Mads Walther-Hansen interview, April-May, 2021). With reference to Figure 16.1, critical thinking is something that is ever present, or should be, within all activities and elements. It is a necessary activity that transforms pedagogical approaches that exemplify *how to do* something to include *why one should do* something and *towards what purpose*. Critical enquiry is an integral element that fuses a student's learning experiences together to develop an understanding of practice. Therefore, educators must empower students to start engaging with critical thinking as often and regularly as possible. Yet there are challenges in this area.

> The challenge is that I think we often prescribe what we mean by critical thinking.... Ostensibly critical thinking is sort of open ended: we're not telling you what to think, we're telling you how to think. But in telling people how to think, we are often telling them what to think.
>
> (Alan Williams interview, April-May, 2021)

Alan went on to discuss a course in one of his institution's programs called "Critical Listening" and he suggested that this was a good analogy where critical listening equates to critical thinking within the program.

> But what are we teaching in critical listening? I think Will Moylan's design is meant to be very holistic, it's meant to include standard

> music theory, frequency spectrum, audible processing and it does a pretty good job of that. But! It also comes with the shackles of Will Moylan's understanding of what it is to listen to music critically and the kinds of music he listens to, or doesn't listen to, and how that applies. So, it's a challenge. I also think that we're very nervous, at least in our program, of going too far in the other direction where it becomes so loose that, "hey any idea you have is valid, it's creative". Or "um, that's not a mistake, you meant to do that". I mean, there's a degree to which you kind of have to hold students accountable. We can't simply be "everything is fine" because that's also not doing them much of a service either. So, I think the hard part for me is: how do you establish sort of bars to set, critical levels of acceptability applied to creativity? You know, I teach a songwriting course. So, what do you do when someone writes what I would call a bad song? I can point out where I think there are some issues, but that's like completely subjective and in fact they might say, "well, I'm not writing that kind of song, I'm writing this kind of song and you don't understand it". And while I might say, "no, what you're actually writing is half-baked, you haven't actually thought it through" and maybe that discussion is there for both of us, but it's an ongoing problem for anybody dealing with creative expression.
>
> (Alan Williams interview, April-May, 2021)

Alan's point is valid: critical thinking, when linked to creative practice, can be a subjective experience, something individual to the listener. I have mentioned previously the way I accommodate this by establishing the function of the song. From this foundation, discussions can ensue regarding whether that function (commercial or expressive) is being met. Nyssim Lefford discussed with me how she approaches critical thinking as an individualised process with her students:

> One thing I do, like from day one, the first time I meet our first years, I'm always making them aware that [throughout their education] they're hearing [about] different perspectives. [From day one], I ask them to bring something in to listen to and they talk about what makes it good, like why did they bring this [recording] and then we discuss different aspects. Different people pick out different things and I ask them, "is there a correct interpretation of this? Is it one perspective that matters?" [And we work out] no, there isn't. I mean, there's certain things that we can objectively measure, you know, this is the SPL, we can objectively measure that, right? But everything else is kind of a little bit of a sliding scale … people are responding in different ways to what they hear.... That is the learning, being able to hold all those different perspectives in your mind at once and being able to distinguish your own subjective view from others, not necessarily value your own more, but to say, oh, okay, I had this reaction and that person has that reaction.
>
> (Nyssim Lefford interview, April-May, 2021)

Considering critical thinking in this way opens the door to multiple perspectives on validation. However, I do not think this poses an issue. After all, as Nyssim highlighted, the learning comes from simply being open to recognising the opinions of others as valid, and to allow everyone the opportunity to provide influence. This has primary importance because, in the professional world, listening has expansive subjective connotations – the public consuming music. Consequently, some institutions consider critical thinking as an advanced concept that students are encouraged to encounter in diverse ways. As Simon Zagorski-Thomas outlines:

> I was looking at University websites recently and I noticed [one institution] had a Masters in music where the module titles are basically 'preparation', 'critical thinking', they're not subject specific at all. A music masters and none of the module titles are 'how to play your instrument', they're all about generic skills that are to do with dealing with the practicalities of a situation. You could almost have general discussions about troubleshooting, problem solving, promoting yourself, understanding the mechanisms of an industry.... The thing that we can't do is throw the baby out with the bath water. Because that works for some people, just getting them thinking about critical thinking and creative thinking. But others haven't got the strength of character at eighteen or nineteen to go "oh I see how this applies to the bigger picture". They're saying, "why aren't I being taught to press the buttons?"
>
> (Simon Zagorski-Thomas interview, April-May, 2021)

Simon went on to discuss some of the abstract ways institutions have approached critical/creative thinking in the past, and it became evident why various institutions may either aspire to produce practitioners, thinkers, performers or a combination of all of these. I prefer a grounded approach that is inclusive of all popular musicians from diverse backgrounds. In synergy with Nyssim's discussions, I talk to students about the concept that "we all think about things and hear things in different ways, and this is what makes us us". I then empower students to believe that their thoughts are valid whilst getting them to recognise value in the opinions of others. I work with students (in mentorships) to develop their engagement with critical thinking and that which they feel is most connected to the function of their music. This process facilitates the students' interaction with consumer-based ideals and individualised preferences. These are soft skill considerations that professional practitioners undertake every time they produce music.

As an educator I know my limitations: if students are philosophical enough and capable of traversing more advanced notions of critical thinking, then they can unpack that with someone a lot smarter than me! I would much prefer to listen to them as they discuss their goals and ambitions, reflect on the musical outcome and what they would do different next time. After this, I say "turn it up and let's enjoy the music!" By working with critical thinking in the ways listed above, we can do all possible to develop graduates whom are both practitioners and thinkers and who have developed individualised understandings of the greater practice of PMP.

REFERENCES

Ciobanu, N. (2017). The applicability of active-participatory methods. *International Journal of Humanities Social Sciences and Education, 4*(11). DOI: 10.20431/2349-0381.0411019

Kerrigan, S. M. (2013). Accommodating creative documentary practice within a revised systems model of creativity. *Journal of Media Practice, 14*(2), 111–127. DOI: 10.1386/jmpr.14.2.111_1.

McIntyre, P., Fulton, J., Paton, E., Kerrigan, S., & Meany, M. (2018). *Educating for creativity within higher education: Integration of research into media practice*. London: Palgrave MacMillan.

17

Class targets

In Chapter 16, I presented a pedagogical framework for learning the greater practice of PMP. I will now outline some integral class targets for students and explore the ways my pedagogical approach supports these targets. This chapter will contextualise some of the profession-based elements from previous chapters, exploring their execution and facilitation in the classroom.

UNDERSTANDING THE RELATIONSHIP AND SCOPE OF CRITICAL LISTENING, MUSICALITY AND EMOTIONAL ENGAGEMENT

When my third-year students arrive at their advanced PMP classes each year, many believe they have already developed the skills of critical listening, musicality and emotional engagement. In all honesty, most haven't even scratched the surface; their critical listening is often one-dimensional:

> I think you've just highlighted a great point there Brendan which is what tends to happen, and I probably did it as well, I wanted to focus on my guitar, but I'm a guitarist. I needed to learn to listen out for my guitar part and to be holistic and see the bigger picture.
>
> (Gary Bromham interview, April-May, 2021)

Students often focus their listening on their own talents and interests: singers listen to vocals, songwriters to lyrics and drummers to the groove. Many students (engineers/producers) think that critical listening is all about sonics or mix balance and spatiality. My doctoral studies found that this was a common and alarming statistic that negatively impacted communication practices.

> I wonder how much of [the students' lack of communication] is because they actually can't hear the musical discrepancies. Students had very little understanding of musical listening, which parts were working together. It is as if the students' critical listening abilities are purely focused on sonics.
>
> (EdD Journal entry, Tutorial 1, February 28, 2018)

DOI: 10.4324/9781003055099-21

In my opinion, this bias is often the result of *technostalgic* and *technofascinated* educational approaches that brainwash students into thinking PMP is all about the technology (old and new) and the sonic signatures they produce. I have alluded previously to the negative impact that this approach can have on matters of equity, inclusiveness and diversity in the classroom. Contrastingly, PMP is all about the song, the music, the way people work together to target aesthetic intentions and craft performances and the emotion you feel when listening to the music. There is a more complex working definition of the term critical listening that needs to be articulated to students so they can learn how to listen in different ways.

In support of this, Sören Johansson suggested that he runs courses where students "develop listening skills and must be able to change listening strategy consciously" (Sören Johansson interview, April-May, 2021). This would mean, for example, that students learn to dial their listening into various elements of the recorded work at will. Students could practice listening and unpacking lyrical messaging and make themselves aware of any emotions they felt because of this. They could then be encouraged to adjust their listening to focus on the musicality of individual elements, if parts have been double tracked, quantised or tuned before focussing on the frequency response or compression of individual elements and the entire mix. Joe Bennett discussed a similar concept:

> Listening lenses is what I call it in critical analysis classes, where we focus on one element of the recording: "everybody just listen to the hi-hat, and what can you take away from it, how is it EQ'd? What grooves is it playing?" Then we'll take another pass, and this time listen to lyric imagery. And you can apply it to any element or sub-element – melody, form, tempo, frequency content. If you take this holistic approach, that the song embodies everything, all parts of it are creatively valid in analysis, because they are all potentially equally important as part of the emotional impact of the creative whole.
>
> (Joe Bennett interview, April-May, 2021)

My research has found final-year students needed to engage with critical listening, musicality and emotional engagement until they had an awakening of the depth of scope to these skills. The subsequent self-realisation – "oh I've only reached this level" – could inspire more growth in different areas (Elliot, EdD interview 2, April 24, 2018). The students of my research perceived that the learning of PMP skills (as outlined in Chapter 1) was represented through a deeper understanding of the skills as compared to knowing how to do them. Students began to understand the complexity of critical listening when they were encouraged to listen in ways that were foreign to them and whilst they undertook many different creative practices. These included pre-production, where listening to references informed their aural library and original demos inspired envisioning; collaborative songwriting, where lyrical phrasing was important for melodic design; and mixing, where musical automation rides did much to increase the perceived dynamic movement and emotion of the song's production. Musicality is intrinsically linked to critical listening; the diversity of musical performance undertaken throughout the pedagogical delivery (diverse music production cultures) immersed students in a context that developed their understanding of performing, listening and working with musicality in many stylistic ways. When one also considers the added complexity of unpacking emotional responses that were elicited from listening, the depth of these engagements can be realised by students nearing the completion of their studies.

SOFT SKILL DEVELOPMENT

Soft skill development is an integral element of PMPP. It includes processes like decision making, communication, time management and artistic touches, and all of these develop over time. Like most young people, I was not good at some of these skills and did not develop them until after university. I learnt most of them tacitly as a part of my post-university apprenticeship at a large studio. I discussed this with Joe Bennett:

> I'm sure you [the interviewer] developed those soft skills yourself as a young professional working in a recording studio many years ago. You felt what it was like to come in late to work with a hangover a couple of times.... Learning about that mismatch between your social life and your professional life, what those tensions are, it's a part of growing up. The soft skill of punctuality – it's one of the most boring skills in the world but, without it, nobody is a professional (and yes, I do quote from the introduction to *This Is Spinal Tap*). Students mess that up, and, fine, we all mess that stuff up when we're eighteen. So I agree with your premise that curriculum design needs to allow soft skills to develop, iteratively and intuitively.
>
> (Joe Bennett interview, April-May, 2021)

Soft skill development needs to be actioned gently when students first arrive at university. I often inform students that I want to develop their social skills, for example, but I primarily give them examples of how to talk to one another and then just help them as they try doing that for the next year. As students progress into final years and I have had a chance to develop their trust, I often become more forthright in calling them out on their lack of soft skills. This is exemplified in many of the videos associated with this book. I find that if I have developed a genuine relationship with students then these moments can do much to give them the necessary reasoning to action improvement in problematic areas. For my students, one soft skill that is always in need of attention is the art of communicating and socialising in diverse socio-musical contexts.

Learning to communicate in diverse music production cultures: Critical thinking and being intuitive and instinctual

If students have developed a complex and rigorous listening approach (as described above), then their communication can be informed from many viewpoints. By working in these ways, the adaptability of the students' communication skills is increased; they learn to communicate with an array of people in different situations rather than focussing on a closed selection of communication skills (asking questions, for example). Communication and psychological skills in PMP landscapes become integral tacit abilities and students should be encouraged to engage with these all the time in order for them to become instinctual. My students learnt them as a form of practice from within diverse music production cultures. For example, students experienced the social contexts that a mixing session catalyses, as compared to a lead vocal session; they learnt that these require different psychological approaches to communication. Mixing is often a solitary process that sometimes requires feedback from the band or the record company. In comparison, lead vocal sessions nurture one individual – the singer. The type of music also often influences the social realities of the music production culture: differing lexicons will be used during a Hip Hop lead vocal session as compared to a Heavy Metal lead vocal session. These are reactive to the subject matter of the song but also the idiosyncrasies of the individuals involved: how they talk, what they talk about and the words that they use.

Jay Hodgson discussed his technique of helping students get used to PMP as a service industry:

> One of the things I love to do when I'm running a mixing class is to have somebody else be the client for a track that the student is actually working on because then they can have that feeling of being identified.... It's not done anonymously. They give feedback and they're just watching them lose their minds but it's so that they can get through their heads that if they're gonna be mix engineers, that a lot of their work is going to be service and a good mix is the one that gets approved.
>
> (Jay Hodgson interview, April May, 2021)

Music-making activities should also be designed as fun and enjoyable experiences and therefore students should utilise humour when possible. Humour is a trait that some people find difficult to manifest. If this is the case, students could be empowered to be their authentic selves because this stands the best chance to develop relationships with people. If a person is perceived to be authentic, this helps to form teams or collaborative hubs and, within these, a student's unique sense of humour may be more readily accepted as a natural part of the team.

In profession-based music-making contexts, students will utilise their listening and critical thinking to inform their communicative and psychological practices without realising they are doing it. In this setting, thinking then becomes a tacit process, something that eventually becomes instinctive and intuitive. Some of my students' accomplishments in this area are epitomised by Lennon's comments: "I reckon I'm good at altering … figuring out how they communicate and then kind of lending myself to that so you're not on two different pages" (EdD interview 3, July 9–10, 2018). Through a continual engagement with the socio-musical applications of PMP my students began to accumulate tacit knowledge and abilities in these environments.

Technology, aesthetics and trial and error applications

In the past my students have learnt to utilise their soft skills when they had to consider various technological options and the agency these choices afforded them. For example, students were tasked to choose different technology when they did an acoustic live recording as compared to electronic tracker/top-liner collaborations. The technology that students used and how they used it affected the artistic touches of their music. In this process, hard and soft skills were often interrelated, and students learnt to "manage the interface between music and machine, art and technology, with a sensitivity to musical expression guiding their own technical performance" (Zak, 2001, p. 168).

Students' reflections on practice-based classes suggested it was an engagement with technology, creativity, conflict and decision making that bridged the space between tutorial and professional practice. For example, the students suggested the synth session (see Video 2) felt real because it was self-directed and trial and error based. Students just "did stuff" and were "chatting, throwing ideas around", and, as a result, students felt they were taken seriously resulting in a positive experience (EdD Focus Group 1, May 29, 2018). Subsequently, the students developed a deeper understanding of soft skills, how contingent these skills could be on trial and error and their utility in PMP.

Release strategies

The creative decisions that contemplate commercial and/or self-expressive musical intentions are also a kind of soft skill. These can be difficult for students to traverse initially and experience breeds confidence in this regard. In pedagogical settings, I often tackle this by challenging students to consider their music in reverse order. I ask students to firstly identify the release strategy they will be utilising and be able to rationalise how their music will be consumed by the public. Who do they want their music to connect with? How will they do that? Do they want to generate commerce and, if so, what release strategy do they have in mind to accomplish this? Has this been done before and was it successful? Students can then steer their songwriting and production to target their release strategies and, following public engagement with their work, experience the realities of their commercial or expressive decisions. If these situations are experienced on multiple occasions, as per the framework described in Figure 16.1, then choices will recurrently present themselves and students will have the option to try different approaches and learn from their experiences. These approaches are only applicable if students have a modern understanding of the music industry, so sufficient concurrent learning in music industry skills should occur as a part of the PMP curriculum.

Russ Hepworth Sawyer explained a similar process that he utilises:

> We talk to students [in a] business module in first year that has two halves. The first half is about looking at the history of the music industry and looking at this whole notion that the music industry, as we understand it, ended in the 1990s but really it started in 1977 when Jacques Attali's book 'Noise' was released and that whole idea that we will one day be in the shit we are now in, where everyone has access to everything. So, it ended in 1977, and getting them to understand that [is important]. When there's this common psyche that everyone understands the music industry being one way, and it's just not that [it's problematic]. It's just unbelievable, and people who should know coming into a degree like ours don't know. So, then the second half, [we turn the module around] it's about themselves and getting themselves to think about what kind of business might they create. They can do anything; they can do a drinks company if they want to. We get them to learn that entrepreneurial stuff.... My point being, we get students to look at it and they sit there and start to think "okay, what do we need to do? We need to create music that does this" and you start to see students going "I want to now start doing music for sync and music for film, I want to do music where I'm actually managing the band or I want to do my rock band but I want to look at these other ways of fan engagement, rather than waiting for people to find me on Spotify".
>
> (Russ Hepworth-Sawyer interview, April-May, 2021)

During my discussions with Prince Charles Alexander, he emphasised the importance of a connection to the Billboard Hot 100 and how that can be used to influence students' creative practice decisions (Prince Charles Alexander interview, April-May, 2021). So much of the music that is consumed today is formulaic and, in my opinion, it is our duty as educators to enlighten our students on the characteristics of the music that is currently being consumed. However, that does not mean that students need to be told that this is the only reason to make music, this only normalises inequitable business models. Rather, educators should outline all options and then promote the students' aspirations: do they want this musical work competing in the commercial market? Do they want to design innovative ways of creating commerce outside of streaming platforms and sales, or use PMP as an expressive avocation? These notions can then be considered at the conceptualisation stage when the enactment of envisioning begins so that a student's actions are tailored towards a release strategy of some kind. Even if that is for personal (self-expressive) use only, as long as there is a purpose, students can be challenged to work towards that purpose.

Profanity

Profanity is common in popular music. As such, it acts as another aesthetic choice students can consider. Previously, I discussed the use of profanity in popular music and popular music culture, and I supported music-making scenarios that use this language when appropriate. This stance comes from a viewpoint that suggests that profanity is widely accepted as a commercial commodity in the music of today and is deeply ingrained in some popular music cultures: Hip Hop, Folk and R&B, for example. It is important for students to consider the collateral realities of using profanity in their music and how it will affect their identity as artists and their release strategies:

> Some genres, if you take out the language and grittiness etc., then you are left with something entirely different – it loses its meaning, its context, its artistry … [However,] this is not to say that we should have a laissez-faire approach to what we can and can't sing or perform in educational settings. Musicianship comes with a sort of responsibility and accountability, so we need to be asking those questions of 'accountable to whom?' Students should learn to be mindful of how they use language and music and their effect, and why they make decisions in certain contexts and not others, and how that constructs their musical identity and intersects with other identities.
>
> (Alexis Kallio interview, April-May, 2021)

Prince Charles Alexander also discussed this:

> I actually don't encourage them to use the language. I am always talking with them about not using the language, even though I know that that's not gonna help them to be competitive.… It's just that you have to be so much more brilliant, so much more insightful … and pulling the human condition into your work in ways that other people just couldn't even have imagined could be done [If you don't use profanity]. Now, everybody's not a Charlie Parker. But to even strive toward this is the beginning of creating other revelations along the way. And if worst comes to worse, if you're like, "you know what … I just can't do it anymore. I'm just gonna have to pop out some fucks", then go for it!
>
> Like, right now in R&B there's a lot of profanity. In pop everybody's "fuck bitch", everybody's doing that, right? So, how can a music student try to be competitive without profanity in a world where the only records that are being distributed are profane records? This is an equation that I don't have to figure out anymore, but my 20-year-old students have to figure that one out. So, I bring them present to it in my classes.
>
> (Prince Charles Alexander interview, April-May, 2021)

By bringing students present to these situations, Prince Charles is encouraging students to engage their critical enquiry to consider the ramifications of their creative decisions (again, the development of soft skills). Critical enquiry of this type promotes the development of a student's innovative capabilities, the forging of new sounds, genres and types of music. Students not only consider that which is deemed as commercially viable – in this case the use of profanity – they consider if they can service the industry in a new and innovative way:

> A lot of the students don't realise how big a moment that is when we're in the studio and we have to make that decision to clean a record [replace words that are considered expletives]. You know, there has to be some thought about servicing as wide an audience as possible. In the spirit of free speech it's like, "okay, that's what you wanna say, that's what you feel is competitive, okay. But now as a CEO of a company I have to tell you, I can't service that on CBS. I can't service that on NBC". You know, I try to put them in that situation.
>
> (Prince Charles Alexander interview, April-May, 2021)

USING REFERENCES

I am a strong believer in the use of songwriting, production and mix references. Students who study with me become accustomed to employing musical references for the entirety of their university degree. One of the greatest issues I see in modern music-making practice is how much of it occurs locked away in home studios by artists working in their own musical bubble. Many talented popular musicians come into the courses I teach but most of them can only work with one type of music and in a certain way, and as a result they are creatively one-dimensional. Oftentimes the music they are creating is of a high standard, but in reality, once you have heard four or five of their songs, you can hear they are using the same tried and tested techniques over and over again, evidence that they may be weak at working in diverse ways. This does not breed flexible music makers. This trait can exist in students of all types: programmers and DJs only working in one genre; vocalists who only sing one style or use the same vocal melisma over and over and session musicians who can't read charts, improvise and play by ear. This is one of the main reasons I feel all popular musicians should be encouraged to expand their skillset and experiment with electronic, acoustic and online music making. The way we can entice students to do so is through listening. Listening to references can inspire students to get outside of their comfort zone and try new things.

For example, I often play my students an array of music that was produced between the 1940s and 1960s and then engage them in a sound-alike session where they use the Les Paul overdubbing[1] technique and an analogue two track tape machine (no computer in sight). The very nature of this sound-on-sound application is new to students and facilitates a very different music-making experience where the influence of diverse technologies and the creative practices they afford can be realised. Students can then be encouraged to autonomously listen to many songs and research the creative practices used on these to further increase their PMP domain knowledge. Through listening to references, students develop a catalogue of information that can inform their agency and their discussions with fellow collaborators whilst making music. Prince Charles Alexander discussed how he uses references as *data points* to inform creative conflicts:

> This is how I tell my students to mitigate the issues and the conflict that can happen. Increase your data points. So, somebody says, "oh, I wanna do this". You don't say, "well, I wanna do that". You say, "well Billy Eilish did this or Billy Idol from 1985 did this". You know, you increase your data points so you're not speaking from a reference of "this is what I like", but you're like "these are the people who are just as intelligent as us that have tried to do this thing before us, and these are the ones that actually did it well … so, how do we measure up against them?" So, I take me out of it. I'm not even in the picture and they're, "listen, well I don't wanna be Billie Idol and I don't wanna be Billie Eilish!" I'm like, "okay. So, you're basically telling me you of all the 60 000 people that just put out a record yesterday, that you are Pablo Picasso?"
>
> (Prince Charles Alexander interview, April-May, 2021)

Musical references not only influence a student's practice, but references (and the aural library they help to develop) are a great source of inspiration for creative conflicts and critical thinking discussions. I discussed critical thinking in the previous chapter; by listening to references and then discussing with others what they liked or did not like, a student can develop an appreciation for music that goes beyond their own personal preferences.

SONGWRITING

One of the most confronting results of the data collection of this book was when I discovered how many institutions are facilitating an education in PMP without providing any formal instruction in songwriting. The term "songwriting" has a very different meaning to "composition" which has traditionalised Western art music connotations. I have already covered why this happens; students are studying science or classical music degrees, etc. The problem is, in my opinion, that you cannot produce music without considering how the musical, rhythmical and lyrical elements of the song influence the production process. A connection to the song is integral from ideation activities through to the mixing process. If practitioners are not doing this, they may be facilitating a recording and not producing music that targets the explicit function of the song. Songwriting and production are interrelated. Therefore, in an ideal world I would promote approaches that engage students with some kind of understanding of songwriting. Joe Bennett discusses this:

> Well, I think we can still develop different skills in different parts of the curriculum, like you don't want to be sitting at a DAW compressing your kick drum while you're trying to understand lyric metaphors or use of imagery. So I think there's an argument for separating out lyric,

> melody, harmonic knowledge, arranging and so on, across assignments, and even into individual courses [modules]. In fact, at Berklee we have explicitly done this for many years – Lyric Writing 1 and Songwriting 1 are separate courses [modules].
>
> (Joe Bennett interview, April-May, 2021)

However, there are other ways to engage students who are not classified as songwriters to work with songwriting.

Using technology as a compositional tool, song fixing and knowing your strengths

If students are not songwriters, they can be introduced to concepts that integrate the songwriting process into their PMP activities. Sam Bennett suggested:

> I always champion the studio as a compositional tool … I think it has massive potential … and it is really important to make it clear to students that there are whole genres that are built on production and it's not really possible to separate the writing and production process. It is like a symbiosis. That's been the case ever since Atari ST [computers] and the first versions of Cubase and Logic and how intrinsic they were to the development of Hip Hop and sampling and the development of dance music. So, to that end, you have entire genres of popular music, Hip Hop being the most 'popular' popular music that are essentially production-based.
>
> (Samantha Bennett interview, April-May, 2021)

The music technology students who are enrolled in my courses do not need to audition to enter their degree and there is no requirement for songwriting. Samantha's approach applies to students like this. The compositional qualities of audio technology now pave the way for the morphing of songwriting and production into a combined application. Similarly, there is added value in encouraging top-line songwriters to consider the affordances of technology as a diversification of their songwriting so that their music is mediated in different ways. Again, this promotes student flexibility. Prince Charles Alexander had a further idea:

> So, my students don't have to know how to songwrite, but what I teach them is how to "song fix", so they don't have to come up with the content, but they might be able to help the content, to connect better with audiences.
>
> (Prince Charles Alexander interview, April-May, 2021)

Song fixing can entail elements that include working with song structure and instrumentation, whether additional sections like breakdowns or key changes are useful, and fundamental concepts of tempo and key are important to ingrain in students' creative practice. The concept of *song length* is always a fun discussion because students can support both consumer (radio) and aesthetic viewpoints. When working with students, I ask them "when you listen to a song (without looking at your watch), how much time do you think went by? How long does it feel?" Often if a song feels shorter than it is, that's okay: it makes you want to play the song again. However, if a song feels six minutes long but is only four and a half minutes then "Houston we have a problem!" A song's *semiotic potentialities* can also be discussed and actioned. Through this lens, students can work on ways of portraying a song's meaning to the listener through sonic, rhythmic and melodic representation.

In profession-based contexts however it may be important to face the reality that some people are just more engaged with, and better at, songwriting. In these situations, it is beneficial if students demonstrate a knowledge of songwriting without engaging in the act of songwriting if others are more skilled. Students must learn to play to their strengths for the benefit of the socio-musical environment.

ANDY: In my production classes I know some students aren't songwriters, so I'll counsel them to play to their strengths.... With these students, you're probably not going to step in and help artists tweak their lyrics, you don't have the credibility even if you have some good ideas. But at the same time, you can try to understand what the artist is trying to say because that develops respect. The artist wants to know that you're listening to the lyrics, and you want to think about the lyrics from a listener's perspective and maybe that helps you understand the artist's intention and where the project is going.

BRENDAN: Students need to understand what they can do, leave their ego at the door and allow others to apply actions in the areas they have strengths.

ANDY: Exactly. Understand what they can offer and be very clear about that when they interact with artists.

(Andy Edelstein interview, April-May, 2021)

MISTAKES ARE OKAY

Many of the educators I talked to suggested that, as a result of their school activities (and perhaps other life events), many students come to university absolutely petrified of failure:

NYSSIM: I think we have … to understand that their [the student's] world is really different than ours. I mean, for one thing there's this underlying terror, and I'm not using that term lightly. I encounter students all the time that are just frightened to death of making a mistake … particularly a social mistake.

BRENDAN: Why is that?

NYSSIM: There are no doubt many factors. As a teacher it leaves me wondering "what if I push that person too far?" Also, I tend to have a

> policy that my students should use their time in school to be bold and if they screw up magnificently, good! Go for it! This is the place to screw up and just learn from it, right? But how can a student take advantage of that freedom to try if they are risk adverse?
>
> (Nyssim Lefford interview, April-May, 2021)

Alexis Kallio discussed the concept of failure and how it can promote student ownership of learning:

> I want students to write their own essay question and they say "but how do I know if it is the right one? How do I know if I am doing it right?" I want to set up a system where students can ask questions, go on a process of enquiry, do some research and take all the tools they have and learn how to use them and completely fail, but get a great grade. They are so scared to fail. We need to equip students with a positive concept of failure and help them take ownership of their learning.
>
> (*Alexis Kallio* interview, April-May, 2021)

I work incredibly hard to let students know that mistakes are okay and, in fact, an essential part of learning as well as a part of professional practice (see Video 23 on the website at www.routledge.com/9780367517632, 1:38–2:28). The only time making a mistake is a bad thing is when a person's ego gets in the way and stops them from owning the mistake and saying: "well that sucks. Let's try something else!" Students from my doctoral study unpacked this as a part of the interview process:

ELLIOT: Out in the real world … there is a lot of lack of forgiveness if [clicks fingers] you kind of need to make a decision, make it quick and make it work … and you know, that is what makes a learning environment safe, I guess, is that you can make those mistakes here.

RESEARCHER: Do you think people in the real world don't make mistakes?

ELLIOT: No, they, … do but, you know … that is another interesting thing because they are clearly allowed to make mistakes.

RESEARCHER: So … what does the real world show us that we could bring into the learning context?

ELLIOT: Well, I guess that mistakes are okay.

(Elliot EdD interview 1, February 26–27, 2018)

Making mistakes is an integral part of PMPP. Through reflection, mistakes help students to develop an understanding of the trial-and-error nature of music making. Bryden Stillie supports this:

> I definitely think that giving students the opportunity to try something, fail at it, but then reflect on why it didn't work, is fundamental to ensure they're getting good quality experiences.
>
> (Bryden Stillie interview, April-May, 2021)

COGNITION AND CREATIVITY IN PMP

Students must be encouraged to interact with the cognitive processes of PMP, for example, envisioning. These processes inform typologies of practice (producer archetypes) and the manifestation stage of music making. Without the cognitive process of music production, there is no road map to success. Music is often made in haphazard ways but getting a bunch of students into a room to just try things with no direction has little utility. There should be some kind of cognitive vision that informs music making, or some form of musical inspiration, like a reference track, that the musician has heard before. Cognition links directly to the practice of PMP creativity where one's aural library informs envisioning sessions. I have used these concepts in my classes and they help steer student activities towards some sort of identifiable goal. An example of my explanation of envisioning to students in a tutorial is available in Video 25 (see Video 25 on the website at www.routledge.com/9780367517632, 11:30–12:43).

The producer's vision, aural library and producer archetypes (auteur, innovator and craftsperson) are specific cognitive concepts and typologies used in my tutorials to promote trajectories of learning. Students should be encouraged to use these in class and during autonomous work to inform their development of tacit skills. Once my students were introduced to these, many suggested they infiltrated their external music-making activities. Georgio stated, "I use those terms in any creative conversation that I'm having, whether it's in a session or out of a session or just talking about concepts" (EdD interview 4, September 24–26, 2018). These three concepts represented an explanation of the cognitive processing necessary in PMP and allowed students to experience how many producers engage with music production creativity in the professional world.

UNDERSTANDING WHO TO MEASURE YOURSELF AGAINST AND BEING A PRO

I love being a respectful reality check for my students. In my classes, I often pose the question "who is your competition?" Now, this may be an Australian thing, but most of the time I get answers that point towards fellow students or, at best, the latest graduate who may have received some kind of local recognition for a single that leads to temporary localised fame for a fleeting moment. To this I reply

> nope! If you're a singer your competition is Mariah, Gaga, Ariana, Charlie Puth, Freddy Mercury, Stevie Wonder, Billie Eilish, Dave Grohl. You bass players and guitarist only have to consider Tal Wilkenfeld, Flea, Satriani, Eddie Van Halen and Steve Vai. Producer/Engineers/mixers think of Max Martin, Steve Albini, Peggy McCreary, Tony Maserati, Bob Clearmountain, Trina Shoemaker, Chris Thomas, Chris and Tom Lorde-Alge, Marcella Araica, oh and ME!

You can pretty much imagine the look of horror on their faces as the reality of being a professional as compared to a hobbyist is confronted.

Most students have suggested to me they want HE to feel real, so let's make it real for them by helping them to rationalise who they are really competing against and how hard those people have worked to be the best. The reasoning behind this approach is that, should they aspire to commercial success, we should empower students to become accountable for their futures and begin thinking that they can become competitive with these people and take action to make it happen.

Time to make it happen!

In 1989 I was the assistant engineer on Midnight Oil's (1990) album *Blue Sky Mining* (and worked closely with producer Warne Livesey and engineer David Nicholas. "Chippa" (Dave), as we used to call him, was amongst the best engineer/mixers I have ever worked with, alongside Chris Lorde Alge and Kevin Shirley. During the mixing stages of *Blue Sky Mining*, Chippa got so sick with the flu that he couldn't work. I remember the day I walked in and Warne said to me, "Chippa's sick so you're up, go get 'King of the Mountain' (one of the songs) crankin' for us. I will come and have a listen in 40 minutes". I still remember the cold shivers on the back of my neck and the absolute fear I was gripped with. Fight or flight kicked in and I wanted to run! Negative thoughts were pulsing through my mind: "I can't do this; I am not ready, I haven't even been here a year! Uni taught me fuck all! What the fuck am I gonna do now!?" I also remember the owner of Rhino, Andrew Scott, coming up to speak with me. Andrew was a big dude, well over 6 foot, and a good friend of Chippa's. He towered over me, leant down and said "listen, I want you to go in there and kick Chippa's arse, okay! Time to make it happen!" he proclaimed as he turned and walked away. It was like a boxer's coach telling him to go and knock the champ out. Quotes from Rocky came flooding back to me. "He's just a man; be more of a man than him! Eye of the tiger, eye of the tiger!" Anyway, I mixed a bit of *King of the Mountain,* it sucked for a while so Warne helped me fix it, then Chippa came back, said it sounded good and started the mix again! In retrospect this was an amazing learning journey for me. All that time I was working at Rhino, I perceived Chippa as above me. That's where the learning comes: I just hadn't worked as hard as him, did not have his experience and did not believe in myself like he did. Subsequently, I went and made all that happen. I am definitely one who prefers actions as compared to words but so that you know: I would mix against anyone now and have had mixes accepted from record companies that use the biggest third-party mixers on the planet. I use this story regularly in my classes to help students understand that, if they aspire towards a commercial career, eventually they are going to have to be prepared to do the work, develop their trade and compete against the best. Make it happen.

WORKING WITH INDUSTRY PROFESSIONALS

I like to provide students an exemplar against which they can measure their readiness for transition into the music industry. This can take place, in one instance, when external professionals (session players or producer/

programmer/sound engineers) are invited into the music-making experience and students can employ their learnt skills whilst working with these individuals. Within my doctoral research there were three classes that involved professional musicians. The students can be seen jamming, discussing their parts, and communicating as if the session was a professional situation. Video 24 P shows this in action on multiple occasions (see Video 24 on the website at www.routledge.com/9780367517632). This is a different scenario as compared to inviting industry practitioners to discuss or exemplify their practice with students who could then copy them. Students working alongside professionals provide the former a rich learning experience. This, in essence, replicates the embodied learning of the apprentice model that was successful in the development of practitioners, and student accountability and communication practices improve when professionals are added to the pedagogical experience.

I would also like to identify that the three examples in Video 24 are all acoustic music-making scenarios. This is because the musical aspirations of many students that year gravitated towards indie/folk-style music. In other course iterations, students have interacted with professional electronic musicians (Rap, Hip Hop, EDM) and the concept of engaging professional practitioners within class-based sessions worked well in diverse styles/genres of music.

By including external professionals in the music-making activities, the students are confronted with some of the actualities of a professional production space. In this scenario students make decisions and react to music-making situations. Elliot reflects on how beneficial this approach was:

It's almost like that the session with the double bass player [professional musician] was kind of like the ultimate test in a way because … it's like there's someone from the outside of the course, who's coming, it's a test. It's a test of whether we're actually professional or not (EdD interview 4, September 24–26, 2018).

These types of applications are not limited to music-making contexts: professional practitioners from diverse areas of the industry – record companies, radio stations and publishing companies, for example – can be included as active participants in creative meetings for student projects, and this facilitates profession-based opportunities for learning to occur.

THE VIBE, EMOTIONAL ENGAGEMENT, COMMUNICATION AND PSYCHOLOGY

Anyone who has made a record of note will tell you how important the vibe is. Now, the vibe, as I explained earlier, can be nurtured in various ways: social interactions, humour, lighting, etc. The thing we must ensure in the classroom is that students do it! Once my students know it's okay to nurture the vibe for classes or their autonomous work, the accountable ones do it all the time. Sometimes my students say to me "oh, I didn't turn the lights down because it's a class". "No!" I exclaim to them, "even when it is a class, it's the real thing". In the videos associated with this book you will see evidence of students nurturing the vibe all the time. To this end they use fairy lights, dim the main lights, communicate like professionals, use humour and, if they feel it is appropriate, they use swear words as positive declarations of musical achievements. They clearly have an understanding of the different ways that music can be manifested, and they utilise the vibe in the design of their music production cultures so as to promote their intended outcomes.

Nurturing the vibe also means that students need to be selective regarding the people with whom they make music. There is nothing more damaging to a session than a bad attitude. But having bad attitudes in class provides useful learning opportunities for students. As Lennon suggested, "I learned that I didn't like AB's, work ethic…. She puts up a very negative front … it's very passive-aggressive, it ruined the whole vibe for me" (EdD interview 2, April 24, 2018). In classes I often talk to students about trying to go into *other people's vibe* (see Video 25 on the website at www.routledge.com/9780367517632, 3:30–5:00). I encourage students to read people's body language and open their senses up to try and feel what a person is putting out. Feeling and experiencing that helps students to react in ways that promote them working with that person effectively. As Elliot suggested, "You can be a little more analytical about how people are acting and that can kind of helps 'cause … then that influences your mood" (EdD interview 2, April 24, 2018). Elliot is suggesting that he learnt to read peoples' headspace and then adjusts his own mood accordingly to nurture the communicative process. Minchella (2018) discusses the concept of *embodied emotional space*. This has great relevance in the field of PMP. It seems some of my students have been able to become aware of their and other peoples' headspace and how best to work within the confines of these situations. These are advanced social activities that I suggest you employ to inspire your students in their PMP classes.

Engagement with psychological approaches through efficient and effective communicative processes can also nurture the vibe during music-making scenarios. Hayley suggested that if there were no social inhibitions in the session, communication comes freely, and collegiality is promoted:

> Even just collectively as a group … any issues that I had with anyone where there was that lack of communication … I guess it's like that whole leaving your ego at the door, like that's definitely become a massive part of it and just we've become more of a team.
>
> (EdD interview 2, April 24, 2018)

Once communication is opened, this provides the opportunity for students to engage with emotional responses whilst making music. Howlett (2007) suggests that music production involves an element of continual connection to emotional transference within the music recording. He suggests that "trusting the emotional response is a quintessential function" (p. 3). There are obvious difficulties in pedagogical contexts when contemplating these processes; the promotion of diverse lexicons in classes, peer pressure and subsequent self-doubt of the students all need to be considered. These can be exacerbated by the influence of the greater university culture, and, in the past, my students have been fearful of what others may think of them if they cry whilst performing, if they swear, if they show happiness, anger or exude sexual attributes.

I approach these situations proactively, but I do so with care. This is done through the promotion of a team mentality where the betterment of the song is at the forefront of all considerations. I find that as students mature into their later years of study they begin to realise the importance of an emotional connection with music, and, through mentorship, can be empowered to feel and express emotion when playing, performing, talking or listening to their team members. This then flows over into their autonomous work. Past students of mine have had realisations of practice generated from these moments: "well, the connection, the emotional connection with your artist you are working with … I think I need to just be gauging people, different people's personalities … and going, 'oh, okay. That's their deal'" (Georgio: EdD: interview 1, February 26–27, 2018).

Emotional connection is an important target within PMP pedagogical approaches. An example of how I help students understand this so they will engage with it regularly is provided at Video 25 (see Video 25 on the website at www.routledge.com/9780367517632, 7:25–11:35). In this example, I asked a student to read the lyrics of her song so everyone could get an idea of the song's emotional targets. When the student read her lyrics, she was embarrassed and read them in a way that was not helpful to the class. This was an explicit case of the negative influence that peer pressure and the greater university context can have on PMPP. I needed to react instinctively to this pedagogical moment, so I then asked her to sing the lyrics instead. When the student did this an emotional connection was made with us all and the students all suggested metaphors they felt resonated with the song's message. Walther-Hansen (2020) outlines the positive influence that metaphors can have when making sense of music and communicating. He states, "we can … think of metaphors as tools for performing musical analyses, taking listening tests, or simply obtaining greater self-awareness about why a music track affected us in a particular way" (p. 5). This activity promoted an authentic emotional connection between classmates and the song. However, the only reason the student felt comfortable singing to the class in this way was because I had developed a professional relationship with her, and subsequently, she respected me and trusted my facilitation.

The tutorial from Video 25 also engaged students with envisioning and pre-production where the interrelationship of psychology, communication and emotional connection was exemplified. Facilitating tutorials of this type catalysed a situation where students' engagement with emotion during their music making became heightened:

ELLIOT: I'm trying to get that emotional like engagement with my music … draw out the emotion from the music sort of deal …

HAYLEY: Emotional engagement is there always and like I guess ties to your communication more than just being a separate thing.

(EdD interview 2, April 24, 2018)

Hayley makes a very good point: in order to understand the people, the song and its artistic intentions, and to connect to emotion, she feels she needs good communication skills. This communication practice, in turn, helped Hayley to nurture the appropriate music production culture that is required to manifest a suitable musical output. Songwriting also intertwines with emotional engagement. By writing, producing, recording, mixing, and critiquing many songs from diverse music production cultures, students experienced the depth of practice that is associated with songwriting and were prompted to engage with emotional transference on many different levels. Students began to understand that emotional engagement is ubiquitous during music production and goes hand in hand with the cultures of music making.

LEARNING TO PRODUCE MUSIC AS OPPOSED TO FACILITATING A RECORDING

Some of the most frustrating class engagements I have had with students involved very talented individuals who work with technology but forget to work with people. Many students program their beats, set up their mics for the vocalist, hit record and then jump on social media while the singer sings! Once the vocalist has done a few takes they will then (after getting off their phones) have a listen and edit playlists into a performance. The problem here is that there is little interaction or care for the socio-musical practices of PMP where performances are generated by people listening, communicating and working together. Students must be empowered to use the soft skills that are associated with the socio-musical interactions of PMP, or I would suggest they are often facilitating recordings and not producing music.

My pedagogical approach aspires towards the development of music producers by helping each student obtain a nuanced understanding of the three pillars of music production outlined in Chapter 2: (1) surveying the musical landscape and designing music production cultures. (2) Making music in different ways and engaging with diverse music production cultures. (3) Explicit application of PMP skills. An example of the positive impact of these three concepts on student learning was presented in my doctoral thesis (see Anthony, 2019, pp. 225–227). The takeaway from this example was that, through an engagement with diverse collaborative practice-based classes that included critical enquiry, students learnt *why* they should produce music in different ways and *how* to apply their practice in diverse contexts. When this approach was applied to students' autonomous work, their agency as producers was informed from tacit knowledge of PMP. These three concepts outline an effective process that higher education teachers can utilise to educate students in the process of producing music.

SUMMARY

The class targets I have listed thus far should collectively work together so that students move past learning how to record and move into the more musical, social and creative practice of producing music. Student interaction with cognition, creativity, instincts, intuition, emotional engagement, the socio-musical and psychological aspects of PMP are the very elements that generate high-quality authentic musical works. These approaches blur the roles of engineer, musician, songwriter and producer into the workings of a PMP collaborative hub. This collaborative hub should value, above all else, the musical work and the high-quality performances that are necessarily constituent to this. In my opinion students often get lazy in this regard and choose not to confront these difficult but integral musical interactions because they can use technology later in less confronting situations whilst they try to fix things. We must do all we can to negate this practice.

> Certainly with practical stuff … it's about the performance you get out of people…. Still, a lot of the students think that the job is to be as in control through the technology as possible and that all the decisions are about fixing stuff, editing stuff, they feel that they are in control of that, rather than the musicians, at least partially in control of that. So [we need to] push them towards Steve Albini and his kind of a servant to the musician.
>
> (Simon Zagorski-Thomas interview, April-May, 2021)

To summarise this target, students should act and practice as professionals. Moreover, they should work for the betterment of the music and strive to produce music, not just record it. Then I trust, through their continued work, that students will start achieving some of the essential pedagogical goals required to demonstrate a true understanding of the greater practice of PMP.

REFERENCES

Anthony, B. (2019). *Perspectives on learning popular music production in higher education from "both sides of the glass"* (Doctorate of Education Thesis, Griffith University).

Howlett, M. (2007). *Fixing the volatile studio: Studio vocal performance techniques*. Paper presented at the 3rd Art of Record Production Conference, Queensland University of Technology, Brisbane, Australia. Retrieved from http://eprints.qut.edu.au/33275/

Midnight Oil. (1990). *Blue Sky Mining* [CD]. [Recorded by Midnight Oil]. Sydney, Australia: Columbia.

Minchella, D. (2018). The poetics of space: The role and co-performance of the spatial environment in popular music production. In S. Bennett & E. Bates (Eds.), *Critical approaches to the production of music and sound* (pp. 41–61). New York: Bloomsbury.

Walther-Hansen, M. (2020). *Making sense of recordings: How cognitive processing of recorded sound works*. New York: Oxford University Press.

Zak, A. (2001). *The poetics of rock: Cutting tracks, making records*. Los Angeles, CA: University of California Press.

NOTE

1 *Les Paul overdubbing* is a process where performances are consecutively recorded and stacked on one track of analogue tape. Once a performance is captured to one track it is then bounced to another track while musicians simultaneously play along with additional parts. This process is ongoing and there is no use of the DAW.

18

Assessment

I will not be rigorously unpacking the utility of various assessment items for PMPP in this book; this type of work would require some considerable scope and appropriately designed data collection methods. My research to date has investigated assessment as a part of my pedagogical approach. In this research I collected data (from students and educators) concerning the different types of assessment that students value and why this is the case. In this chapter, I will present some perspectives on suitable assessment approaches for PMPP derived through this experience. Note that this will constitute a limited selection of exemplars and should not be construed as an extensive list.

PMP ASSESSMENT 101

Advanced PMP assessment items require explicit goals that support the overarching learning outcomes of the course but should be flexible so as to also nurture the individualised needs of every student. I am in favour of final-year students taking control of their learning (and their assessment items), as they progress into more advanced concepts. This does not mean students are free to do anything they want; rather, students can be positioned in various assessment frameworks where they can apply critical enquiry as to how they want to engage with an experience and then unpack why they approached the assessment in this way. This style of assessment allows the student to establish ownership of their learning experience. This aligns with the seminal literature including, "situated learning" (Lave Wenger, 1991), "co-creation" (Chemi & Krogh, 2017) and "active participatory learning" (Ciobanu, 2017) that I have discussed earlier. Bryden Stillie supports this approach:

> We're never really constructive about project briefs, I think that's something that enables students to be free and creative when they're working…. We have some things that we want to nail down that they can demonstrate … but actually the reality is whatever the end product is, it's all about them [the student].
>
> (Bryden Stillie interview, April-May, 2021)

DOI: 10.4324/9781003055099-22

Intentions, producing music and reflection on outcomes

Most assessment items that I utilise for advanced PMPP engage students with (1) articulating their intentions, (2) producing music and (3) reflecting on outcomes. The critical enquiry that is applied in these scenarios takes place in both individual and group contexts with peers or industry personnel. I spoke to Alan Williams about this approach:

> Yes, I love that.... That's difficult to do though. I think often our intention is to be more open minded and less directive, but, at the same time, I have found that every time I pose the most open-ended assignments I get blank looks back. Most students are like "no no no, I am paying you money! Teach me stuff".
>
> (Alan Williams interview, April-May, 2021)

I mean, I never said it would be easy! This is why these types of assessment approaches are more-so suited to students who have got a few years of HE under their belt, broken any shackles the schooling system may have placed on their understanding of education and are ready to take control of their learning. Critical enquiry is an integral element of assessment activity because it allows students to not only learn a lot about themselves and how to develop creative practices that work for them, but through group reflection they can develop an understanding of their peer's appreciation for their practice through their music. They discover the ways their work is servicing those who would consume and potentially validate it as authentic. These are learning approaches that help to unpack *why* they should do things a certain way. Andy Edelstein reflects on his approach:

> If a course is working, my students are going to learn more about how their brains work, their instincts, their strengths, what's compelling to them, and what they gravitate towards. Students are required to articulate their goals and plans then critique their own and each other's work. It sounds easy enough but is often quite a challenge. Trying to describe what we're doing and why and attempting to understand other students' motivations reveals biases and discontinuities in our own perspectives and assumptions. 'Why?' is often the hardest question.
>
> (Edelstein, 2021, From Berklee website)

By applying critical enquiry to assessment items students develop reflective capabilities "so they will be able to apply new knowledge to complex situations" (Lebler, 2008, p. 194). Reflection empowers students to be strategic, develop an understanding of their strengths and weaknesses and manage these effectively (Claxton, 1999). Lebler (2008) supports the use of student journals that accompany creative works and reflect on creative practice:

> One of the most common forms of structured reflection is journal writing. The writing of journals enables reflection and the creation of

> ideas and patterns of order out of the mainly random events of experience. Activity and reflection are complementary and mutually supportive; action alone is blind, reflection alone is impotent.
>
> (2008, p. 194)

Journal writing is a very different type of writing as compared to an academic paper. Many participants of my research felt that their university education focussed too much on inconsequential academic writing and not enough PMP activities.

> About 70% of the time you are writing some sort of paper, relating to an issue that you don't really care about and you want to produce and engineer instead of reading about scientific theory and Tia DeNora all of the time!
>
> (International student survey response, May, 2021)

Written work is useful for PMP graduates who aspire towards academic or professional futures, but my research has shown that many students want to do more things and write less. However, critical reflection is integral to deeper learning so there is added value if students can present their thoughts using modern mediums that they find appealing.

Pitch sessions, video journals, web blogs

I am not one that mandates that all popular musicians must show a high degree of writing ability. I am dyslexic, can't spell and it has taken me 30 years to be able to write in any kind of scholarly way. Many of my students are the same and prefer to speak publicly, engage in group discussions or record a video where they describe their intentions, practice and outcomes. I have mentioned before how I engage my students in pitch sessions where they articulate to the class their plan for a proposed musical work. Pitch sessions include any references that inform envisioning, creative practice approaches and reasoning behind these actions due to the function of the song and its lyrical messaging. The use of video journals or weblogs is also something colleagues of mine were in favour of:

> I'm sort of in two minds because you have to be able to write an email … or write in various job [applications] so I think it's important, but a lot of them [students] just don't have the skills. So, what we're gonna do next time around is we're gonna replace a lot of the write-ups with videos because the world is so video based anyway … it's like a new transferrable skill that's become really useful.
>
> (Kirsten Hermes interview, April-May 2021)

The integration of videos or weblogs resonates strongly with the modernised online world that PMP graduates will experience. In my opinion, these can form innovative new territories for students to articulate their critical reflections in PMP assessment items.

RELEASE MATERIAL AND LEARNING JOURNEYS

I utilise two main frameworks for assessment items: (1) release material and (2) learning journeys. The basic difference between these two is, I would hope, obvious: one places importance on the final result and the other does not. Release material can often be associated with a deadline or time limit (one 4-hour session to accomplish the task, for example) and this adds profession-based pressure on students whilst they are working. Student accountability is an important pedagogical goal (and this will be discussed in the following chapter); by getting students to work on projects that will be released to the public, educators can empower students to become more connected to working like professionals. Student accountability increases if students feel they are going to be associated with music that is going to be made publicly available. Students' comments confirm this: "if you wanna release that and use that, then people are gonna see it and you wanna put your best foot forward" (Lennon EdD interview 3, July 9–10, 2018). By associating a release ideology with some assessment items, students develop a sense of pride in the quality of their music. Alan Williams commented on his work in this regard:

> I think that once the students realised that public means public.... That's a very different place than most student projects are going to end up on.... But it pointed out that this was an element that we could be instilling in our students for everything they do. [Namely:] "we are going to put this up publicly, take it seriously, be sure that you are happy with what you are putting forward". Because, whether it is career aspirational or not, I think that most people do believe that whatever they are working on will eventually end up in the public domain. In which case instilling that sense of ownership is a really good thing and something we could be doing better at university.
>
> (Alan Williams interview, April-May, 2021)

Alan and I went on to discuss that there are overarching considerations with release material that must be contemplated by educators since they provide the possibility of feedback from the greater public (via social networks), and, as such, may positively or negatively impact the student experience.

Furthermore, my students appreciated having diverse assessment types because this helps to tailor to individual student aspirations and promote multidisciplinary graduates. As one student explained:

> 'cause [with] real world, there's so many possibilities ... there was a good range of different things [in the tutorials and assessment] and I liked the way that the assessment was very self-driven. It ties to accountability and real-world practice.
>
> (Hayley, EdD interview 4, September 24–26, 2018)

In my courses, it has been important that some assessment activities were learning journeys rather than projects in which the musical outcome was given a mark. I have discussed learning journeys previously where students, using their own volition, find a way through to completion of the project. The assessable part of the project focusses on the student's critical reflections on practice and their learning – the musical work is not marked but submitted for context only. This is a valuable approach because many students are not capable of producing high-quality release-ready material after a few years of learning how to produce music. Again, through critical enquiry and reflection on intentions, practice and outcomes students can utilise learning journeys to reflect on where they are at as practitioners and what they need to work on, and this opens doorways to future projects.

PEER AND SELF-ASSESSMENT

Assessment items where students receive a mark and some feedback from the educator alone, in my opinion, consider student music from a narrow perspective. This approach is more suitable for the arguably less creative applications of sound engineering. Is it important that students hear what the educator thinks of their music? Sure. Is it the most important thing? No! Students agree:

> A student's work should be assessed and judged by a number of people. Having a single educator determine the creative value of your work can be unfair as I have personally experienced lecturers reject my work based on their own musical taste. When delivering my work to fresh ears and other educators, I received different opinions showing that music production is very subjective depending on who's listening. It should be a shared experience.
>
> (International student survey response, May, 2021)

I have found it valuable to include a student's peers' or industry partner's reflections on their work, because, in most cases, this will replicate the current consumption of music. My institution is well known for the application of peer assessment activities (see Lebler, 2008) and I have been involved in this process since the early 2000s. Peer assessment, when facilitated well by educators, is useful for generating a consumer-based reflection of student work. This type of assessment practice also allows for a large scope of musical feedback to be provided to the students. For my advanced PMP classes, I run music portfolio peer panels. In these scenarios, every student's music portfolio (two tracks) is critiqued and assessed by approximately eight of their peers and the educator in an online forum. Every peer panel participant also provides approximately 70 words of constructive feedback for each track that a student submits. The group then marks the student's work in accordance with the assessment criteria and rubric and an average mark (educator/peers) is calculated. The whole process is facilitated and, if necessary, moderated by the educator.

In peer panels, students are never present when their music is marked. However, sometimes students may have worked on the music that is being assessed. In these cases, to satisfy matters of equity, students who have worked on the music can provide context to the group regarding the musical work, but they take no participation in the allocation of marks. Online peer panels also have great utility as pedagogical engagements because not only do they provide a collection of well-informed feedback, they also allow students to critically reflect on their listening experience and present their thoughts on the musical work to the group, and this promotes the development of a student's communication processes. Peer panels also pave the way for future collaborations because often students will be inspired by each other's work and seek new collaborative partners out.

During peer panels there is significant importance placed on the educator to not take over the panel engagement. The educator is there to facilitate and not lead or inform students' thoughts and opinions. Once students have voiced their opinions, the educator can then discuss the music like a peer, and this helps to break down the educator/student dynamic. An example of a peer panel is exemplified in Video 26. Due to copyright restrictions there is no music in this video but the panel engagement and discussions are of premium importance here. This panel is run by a colleague of mine, Caleb James. Caleb is the songwriting course convenor at my institution and we often work together for moderation purposes. I have utilised this video as an exemplar of how staff from varying departments can keep the lines of communication open and work together to target a high-quality student experience (see Video 26 on the website at www.routledge.com/9780367517632).

Self-assessment

Don Lebler writes: "the ability to conduct independent and autonomous assessment of work while it is in production is one of the characteristics of professionalism" (Lebler, 2014, p. 181). Self-assessment activities allow students to critically consider their music, agency and professional practice and develop their practice independently. This process bridges to the real world and is something producers do on a continual basis – measure their work and performance. I have found utility in this process within individual and group contexts; sometimes learning journeys can be better assessed as a self-reflection of practice (as opposed to an educator grading a musical work). The framing of this assessment can be rather diverse including online forms or reflective journals with numerous assessable criteria in which students self-mark, a process which can be moderated by educators should the need arise. One of the most useful outcomes of the self-assessment process is that it fuels honest mentorship discussions with the educator regarding student progression.

MODERN ASSESSMENT, PROJECT MANAGEMENT AND ENTREPRENEURIAL WORKFLOWS

Students value assessment if it relates to the professional world and if it tangibly deepens their understanding of PMP:

> The assessment was totally relevant and just really became practice of what we were doing in class (which was then applied in life

> practice anyway). It also helped to delve into production … to develop understanding in a deeper way.
>
> (EdD Student survey response, 2018)

Different types of assessment earn the approval of students for varying reasons:

> I definitely thought the designed assessment tasks were of value, especially the tracker assessment. I personally, as a singer/songwriter, found it very useful to discuss and explore the tracker assessment as it introduced and allowed me to practise more top-line writing which is something I want to carry on doing after university.
>
> (EdD Student survey response, 2018)

Two assessment items I have utilised in the past are listed here. Both of these items have been met with exceptionally high student experience ratings.

Live recording workshops (for public release)

The live recording workshops are student-led, practice-based, four-hour recording sessions for one of the students' original songs, and all music is recorded completely live. The students also shoot a video clip at the same time. Video editing and mixing is permitted following the session but this activity engages students with a live recording culture that promotes the generation of a collective performance. There is no limitation on the music that is performed; it can be of any genre. Students work in groups of up to eight people and are required to facilitate a pitch session to the class prior to the session. The pitch should cover all aspects of the production process including songwriting, references, musical intentions, target audience instrumentation, performance and session setup (audio). Students are assessed on overall musicality and performance, audio and video content (which is uploaded as a YouTube link) as well as their individual contribution to the project. This item is peer assessed as a part of each student's music portfolio. An example of this assessment item in action can be seen in Video 24 P (see Video 24 on the website at www.routledge.com/9780367517632, 0:00–4:44).

The tracker assessment (learning journey)

The tracker production process is a growing trend in music production where top-line songwriters (lyrics and melody) work with music programmers and music producers. In these music-making cultures, production, songwriting and recording often happen concurrently and the backing track is generated primarily via programming within the DAW. However, the creative process is flexible and engages with face-to-face and remote/online communication and collaboration processes.

Students are introduced to the tracker process via a lecture, after which they are tasked with engaging with the process to generate an original musical work via remote and online collaboration. The musical work is required to be of an electronic music genre of some sort in order to promote that all students experiment with DAW-based production. The students are split into production teams (two or three members) and take part in an initial creative meeting in the tutorials before commencing work outside of the tutorial. In later tutorials, students bring their work in for peer feedback and they are marked via a creative synthesis (a written journal or a video blog) that explores each student's perceptions of engaging with songwriting and production in this way. The students work in production teams but are assessed individually. There is particular emphasis in the assessment item regarding the influence that remote and online communication and collaboration processes had on the creative output and the student's use of technology, in particular the DAW. The musical work itself is not marked because the assessment item is designed as a learning journey and the students' unpacking of their perceptions during the process is key.

Overwhelmingly, students find value in the creative synthesis because, within it, they are able to express their feelings regarding creative practice. By doing so, their learning and understanding of the benefits of the tracker process are nurtured:

> I learned more-so from unpacking every aspect of the task in the journal. I reflected on the language we used, the effects of body language and knowledge of the collaborator. There are so many factors that contribute to the workflow approach and result of the tracker process. Every little bit needs to be considered. And I only knew exactly what I learned when I unpacked it.
>
> (EdD Student response tracker survey, 2018)

Project management

Through my discussions with Sören Johansson, Russ Hepworth Sawyer and Mark Marrington, it has become evident how useful project management can be as an assessable item that prepares students for life after HE. In these situations, students can be paired with industry partners (clients) to liaise and develop a product that targets the client's needs:

> They [students] find it refreshing, I mean, they get used to us [educators] … and these people provide a much-needed contrasting breath of fresh air 'cause they come in and they just talk the talk, really. They are clearly immersed in their everyday life as people that know production. I think the students respond to that. It helps to balance things out a bit [university life/professional life].
>
> (Mark Marrington interview, April-May, 2021)

Project management also directs a student's agency towards a musical work that targets the client's aesthetic requirements (not the student's), but it also makes PMP something that becomes a part of a greater activity: running a business, for example. "Yeah, the idea with my program is that we assess the skills of project management, so all the song writing and all the mixing, you know are side effects…" (Sören Joansson interview, April-May, 2021). In this scenario, students are required to be actively engaging with autonomous entrepreneurial workflows and are marked on their liaisons with clients, how they conducted business, their time management, their budgeting, etc.

The educators with whom I discussed these approaches suggested that their institutions allocate funding for industry partners to become active participants in their pedagogical approaches, and that students provide a service for them. There are obvious limitations to this model because decision makers may balk at employing educators and allocating additional funding for industry partners. However, this does leave institutions in a perfect place to consider innovative and diverse human resource applications. As an example, final courses for PMP students could employ a current record or publishing company executive or marketing manager from an advertising company as a sessional lecturer. This person could supply briefs and projects to students and liaise with them as a client. Toby Seay discussed a model similar to this idea:

> We had a professor who just retired, but he taught the senior capstone of music production, and he didn't teach it in a studio…. So, like, it's budget management, talent, management etc. He had the entire project [mapped out] – "you do this on your own studio time, but here are all the things that you have to contend with. Your studio time is gonna cost this much per hour and you have to fit it into this kind of a budget and you have to provide preproduction and the demos".
>
> (Toby Seay interview, April–May, 2021)

If educators aspire to prepare graduates so that they can integrate into the music industry, whether that be in current or innovative ways, an engagement with project management could be a very useful assessable item to manage this in diverse, flexible and relevant ways.

REFERENCES

Chemi, T., & Krogh, L. (Eds). (2017). *Co-creation in higher education: Students and educators preparing creatively and collaboratively to the challenge of the future*. Rotterdam: Sense.

Ciobanu, N. (2017). The applicability of active-participatory methods. *International Journal of Humanities Social Sciences and Education, 4*(11). DOI: 10.20431/2349-0381.0411019

Claxton, G. (1999). *Wise up: The challenge of lifelong learning*. London: Bloomsbury.

Edelstein, A. (2021). *Andy Edlestein: Associate Professor* (Berklee College of Music online testimonial). Retrieved from https://college.berklee.edu/people/andy-edelstein

Lave, J., & Wenger, E. (1991). *Situated learning: Legitimate peripheral participation*. Cambridge, MA: Cambridge University Press.

Lebler, D. (2008). Popular music pedagogy: Peer learning in practice. *Music Education Research, 10*(2), 193–213. DOI: 10.1080/14613800802079056

Lebler, D. (2014). Promoting professionalism: Developing self-assessment in a popular music program. *Proceedings of the 20th International Seminar of the ISME Commission on the Education of the Professional Musician* (pp. 179–191). Retrieved from https://www.isme.org/events/past

19

Student accountability and educator mentorships

When I considered the scholarly literature, the comments I have collected from international educators and students and the findings from my doctoral research, one major PMP pedagogical goal is evident. Students must become accountable for their learning and development as PMP practitioners. That said, we cannot expect students to become accountable as soon as they enter university or to find it easy to do so when more experienced.

> It's great saying they should be more accountable, and I think this speaks back to the … marketisation of education. You ask a question to the group and there will be silence, because they're waiting for the right answer from you, that is at the back of the book they should look up! [This is the education they are accustomed to from schooling] and you say, "sorry gang, this doesn't work like that". I think as educators, we need a transition change management style approach to develop a culture of accountability for students. We can't just expect eighteen-year olds to be suddenly accountable.
>
> (Paul Thompson interview, April-May, 2021)

Paul went on to suggest that the HE experience is all about enrichment. That students start at one end and come out the other end a better, more rounded human being. To accomplish this, educators need to work with university governance to develop suitable strategies.

> I think the whole problem goes back to the marketization of education idea, the dominant voice in this is about employability and industry. And you're like, "well where the fuck has industry been all these years?" [We need to break] the cultural bonds in which they've been chained up in, because [students think] learning takes place in the classroom, because that's their experience, but their best learning experience is in the park, or at the school disco. Learning actually takes place in lots of different ways aside from that, and university is the same. It's irrelevant what your subject is as long as you have a passion for it. You're going through a process of personal enrichment.
>
> (Paul Thompson interview, April-May, 2021)

DOI: 10.4324/9781003055099-23

In my experience, a student's maturation in terms of accountability is linked to self-awareness, reflexivity and constant adaptation to life. These are difficult spaces for students to traverse alone and, therefore, there is much to be gained through the development of educator mentorships.

MENTORSHIPS

Educators can implement transformational frameworks in the form of mentorships which target the development of student accountability over time. When I spoke to Russ Hepworth-Sawyer and Mark Marrington they highlighted the importance of mentorship-like relationships between the educator and student which, in their case, are actioned through one-to-one pedagogical interactions (Russ Hepworth-Sawyer; Mark Marrington interview, April-May 2021). My institution does not currently support one-to-one tuition for PMP so my approach is more holistic (it is always happening, inside and outside of class activities). The first process I employ in these mentorships is to talk to students with the passion and honesty that is obvious in Paul's words above. Educator authenticity goes a long way in these difficult and socially volatile pedagogical applications. It is unfortunate that in international student surveys I have received numerous comments suggesting some educators were not empathetic and empowering human beings. Allegedly, some acted like egotistical elitists, speak down to students, are rude and show little interest in student progression (Various international student survey responses, May, 2021). Authentic and genuine mentorships have great utility because they can be used to respectfully challenge students in many areas including reflexivity and accountability. Mentorships achieve this by breaking down the educator/student dynamic to that of equals. Once this is done, authentic and genuine conversations may ensue and these can be used to empower the student to continue doing what they are doing well, and to make changes where change is necessary. PMP students value this type of approach since music making is a personal activity that often uncovers an individual's nature as a human being. Music making requires people to engage with emotional experiences and these are oftentimes difficult territories for young people to explore without the guidance of a mentor.

> I have the impression that most other universities have a much more distinct line between teacher and student. I feel that being treated as an equal and a peer by my educator greatly improved my learning experience, and better prepared me to work in the music industry as a producer, where you don't have a teacher there to instruct you. The teaching model in my classes encouraged me to communicate effectively, take control of recording sessions, and push myself to continually try new things and improve my skills as a producer. It made the experience much more of a personal journey, and I feel that many other students miss out on this.
>
> (International student survey response, May, 2021)

Cumming (2000) discusses teachers who extend "their role beyond the transmission of technical skills, or even of formal and stylistic understanding, to include a challenge to emotional life, and rationality, in the name of musical interpretation" (p. 8). As an educator I embrace this role from day one with PMP students and utilise it in different ways as students move through their learning/enrichment experience.

Breaking down the educator/student dynamic

I have discussed the educator/student dynamic extensively in this book, and, in doing so, I have promoted an approach that supports this normative relationship being altered to render educator and student as equals. By doing so, educators set up the framework needed for the mentorship role to evolve. I spoke at length with Alan Williams about developing mentorships:

ALAN: You know, back when I was doing professional engineering and production work, the sense of not feeling confident about what I was doing was pervasive and so, from my perspective, I would have benefitted had somebody helped me to say, you know, what you're doing is actually as good as you hoped that it is.... I think we need to consider the ways we teach people and what we hope that they can get out of it. I think that self-knowledge is a great way to frame ... getting confidence but also to have a realistic sense of where we are with things.

BRENDAN: Is there value in forming professional relationships and mentorships with students so that there's comfort in these communications that need to happen?

ALAN: Well, that's exactly it, and that's what takes time, and it takes a tonne of energy. As an employee, there's a certain amount of time that's expected and a certain amount of time that is allocated. The union that I'm a part of would argue "stop doing the extra things you're doing"!

BRENDAN: But that's the good stuff [laughs]. That's the stuff that counts that many places don't allocate enough into workloads.

ALAN: That's exactly it. I'm devoting way more energy than I should have to the Pink Floyd project [involving students], but I'm also mindful that we've created a particular environment where there is a great deal of trust and comfort level between myself and the students.... Part of that is that they were able to work with some professionals that I brought in to track the session. They happened to be alumni, so students thought: "okay, wow. So, you are someone that's been through the program I have. You are who I want to be in five years. I am studying the way you're interacting. I'm studying how quickly you're getting this done. I'm studying how you relate to Alan". Like we're all kind of absorbing all these things at the same time and clearly what has happened is my students and I see one another as equals in value with different types of experience. I'm not even going to say degrees – it's not hierarchical. It's placement. It's types of experience and understandings and when I say, "Um, something happened in the bridge" that's not saying "you morons! How can you not play this right?" They're like, "yeah. Sorry, I screwed up the chords" and someone can say "you know, we're having a problem with this transition" and usually for me it's just a prompt – something was wrong. I know what was wrong, I'm not gonna say it directly; I want to see what they can do and at this point they do it.

BRENDAN: Yeah!… if they feel like they're your equal, and they see people who were where they are now and they are now professionals, then they get a realisation, "I can actually do this. I'm the same as these people".

ALAN: Absolutely. I am of equal value to everyone in the room. Every potential is there. This is sort of what I am known for in my program for better and for worse. So, the risk there is that you become too overly familiar, that lines get crossed or that somehow the perception is that, well, you love your students so you accept behaviours that frankly aren't acceptable. The other version of that is I try to develop a sense of earning their respect.… "I got no problem with getting down and getting my hands dirty with you guys but understand that I don't have to do it, understand that I've been getting my hands dirty since before you were born. I don't wanna have to keep reminding you of this, but I want you to know it". I think that there is kind of a progressive step in how you approach that. So, from the outset I do make it clear: "hi, I'm an old guy, I have a lot of experience, I've read a lot of books, I've played with a lot of electronic things, I've put out some records, I've done it! So, listen to me 'cause I am the expert!" [laughs]. Within a week or two I wanna get rid of that but I need to kind of establish that early so that the comfort level sort of evolves organically and that by halfway through a semester they've forgotten why they remember I'm the teacher, but it's still clear that I have a role that's a little different from them.

BRENDAN: Fantastic.

ALAN: But when we're immersed in the actual projects, by the second part of it I am essentially their equal in the endeavour…

(Alan Williams interview, April-May, 2021)

By operating how he has suggested, Alan has developed a dynamic with his students where everyone is perceived as equal, but he, as the educator and professional, is known to have experience in the areas that the students are learning, and this garners him some respect. At the same time, Alan will engage in music making with his students and converse with them as equals. This allows students to view themselves as the same as Alan: he is a professional and they can be too. These are powerful pedagogical approaches and they are beneficial in advanced contexts as compared to more traditionalised master/apprentice models.

My students became empowered when similar situations arose in my courses (Anthony, 2019). As one student mused: "it's like you became one of us … everyone's cool, everyone's … putting their ideas forward and I think in the last couple of ones … you've become just part of that culture … which is the real thing" (Georgio, EdD interview 4, September 24–26, 2018). By fostering meaningful professional relationships with students, educators pave the way for mentorships that can span a student's career (yes, even after university), help students with their reflexivity and accountability and encourage them to bridge their mindsets to that of professional practitioners.

Increasing accountability: Student headspace (uni is just uni)

Student accountability is promoted if they value the university experience as helpful towards their future; students should be empowered to steer well clear of the whole *uni is just uni* mantra. Students from my research became accountable when the tutorial activities resembled professional practice and value was placed on the creation of music as opposed to lessons on using technology. Students also valued a diverse range of creative practices that they could integrate into their autonomous music making. These activities included historical approaches to music production (live tracking to two-track analogue tape, for example) as well as modern applications like the tracker process and third-party mixing. Additionally, students valued engaging with every procedural step of music production (regarding both acoustic and electronic music production processes) from the cognitive beginnings of songwriting and pre-production through to recording and mixing. These activities were integrated with lessons on the portfolio musician and the modern music industry to help contextualise the work students were doing and broaden their view of its situation. Mentorships also aided individualised student learning targets and many of them became more accountable because they associated their university life with their future as professional practitioners.

FROM PASSIVENESS TO PRO-ACTIVENESS

One of the most common traits I find in students is that they are often overly passive in class. This is often a by-product of the fact that they are used to being taught. Many educators I interviewed often referred to the regular *stunned mullet faces* they received from students when student-led applications were broached. There are many ways I have tried to remedy this over time, but the most successful method I have found is to confront students about it directly in class. I do this from varying perspectives that include being caring and empathetic, empowering and proactive and sometimes I just sit in class and don't talk until one of the students steps up and engages. This is exemplified in Video 27. Once passiveness has been negated as much as possible, students get on with the making of music, the doing, the learning of tacit and soft skills, and as a result they become more accountable (see Video 27 on the website at www.routledge.com/9780367517632).

ACCOUNTABILITY TO THE GROUP, TO ONESELF AND TO ONE'S FUTURE

If educators can stimulate a team mentality in their classes, something that resembles the collaborative hub that I have talked about previously, then student accountability can be specifically tied to accountability to the group involved in the music making. My research has found that a growing sense of loyalty to the team can be nurtured, and, in my experience, this increased as the classes transitioned through more difficult learning experiences. This concept resonates strongly with professional practice. Producer Dave Jerden declares "there has to be a bond made between the producer, the band, and the engineer. It's almost like a family, and you have to understand each other and establish trust" (as cited in Zak, 2001, p. 163). Students from my classes learnt the importance of the trust Jerden mentions above, and, as a result, many became accountable to each other by the end of the tutorial experiences. However, it was important for students to initially feel what it was like to not have that trust:

> I think that was an evolving process, but it was more exponential at the start of the year ... experiences sort of kicked us in the guts and realities came forth ... I think people just realised that they're not gonna get anything out of it if they don't make themselves accountable....
>
> (Georgio, EdD interview 4, September 24–26, 2018)

The students from my doctoral research became accountable to their working groups by taking responsibility for their roles. Hayley attests to this shift: "I feel like everyone's just sort of realised how important it is to ... work together as a team.... We were just there to record music and do the song justice" (Hayley, EdD interview 4, September 24–26, 2018). Similarly, Adam Bell described to me how he communicates to students about accountability to the student group:

> I say to them: "you need to find an effective way to communicate. You don't wanna be burning your bridges, right? You don't wanna be that jerk in the group. You gotta show up. You've gotta be accountable to them or else you let them down".
>
> (Adam Bell interview, April-May, 2021)

Student individual accountability is therefore encouraged with a subsequent loyalty to the group and the collaborative nature of the learning experience – the two work in synchronicity.

Individual accountability (phones, social media and personalities)

Individual student accountability seems a much more complex issue. This is because each case is intrinsically different and specific to the individual. In open and honest mentorship discussions, the student participants of my research suggested that individual student accountability, in the end, is attributed to a decision made by that individual. This may seem logical but there is much to consider here that influences a student's decision on whether they will be accountable or not. Students possess unique personality traits and, subsequently, mentors can help students with their reflexivity over the course of their study and help them with any weaknesses they may feel they have. Within these mentorships, accountability discussions can be undertaken with students in private settings and personalised pathways to success can be developed. I will go into some of the experiences I have had in this regard a little later in this chapter.

Many of the participants of my research have confirmed that mobile phones and the use of social media are distractions during classes. I absolutely agree and find nothing more frustrating than seeing engineers, singers or musicians posting on Facebook when they should be communicating and focussing on the music. However, mobile phones and similar technology can be useful tools within a session, whether used as a tuner, to take notes or to play musical and sonic references. The use of social media chat groups also benefitted my students by supplying the necessary communicative framework to prepare for upcoming sessions. Social media, in this sense, increased individual student accountability and this helped position student practices within the scope of the modern professional. Therefore, it is important that educators discuss with students how social media can be both beneficial and harmful to their music-making experiences at university, and how the development of professional, efficient and effective workflows will promote longevity in their selected paths into the future.

How you practice is how you play: Creative conflicts and ego

In previous chapters I alluded to the importance of the *how you practice is how you play* mantra (see Video 23 on the website at www.routledge.com/9780367517632, 7:30–8:35). In this video example, I exemplified to students how Tiger Woods was as accountable when he practised on the driving range every time he hit a ball (whist no-one was watching) as when he was playing at the US Masters (I know, another sports analogy). The rationale of this approach for PMP is that if students thought this way, they would always do their best as compared to just when a project was assessable or when the project may be released. During one of my doctoral focus group discussions, I played students some videos of their classes and they watched as the music-making experience was a failure because students were not engaging, not rehearsing and not being accountable.

> There's that saying, 'how you practise is how you play'... and that was a massive thing in all of this production stuff ... the way that we are treating class [should be as if it was] real life ... I feel like sometimes when that was lacking, it really affected everything. It showed, and we all felt that sense of unprofessionalism.
>
> (Hayley, EdD Focus Group 1, May 29, 2018)

How you practice is how you play confronted my students to ask themselves, "do I want to be a professional or do I want to be a hobbyist?" All students were encouraged (by the mentor) to bring professional attitudes to classes and to engage like professionals, but this was firmly positioned as the responsibility of the student. Utilising mantras like this aid student accountability. My advice to educators is to use any mantras that have been helpful to them since this helps their advice feel authentic to students. Some students reacted well to this technique, others did not, and that in and of itself is a learning experience for students. Some students realised that they were just not cut out to be a PMP professional and, if that is the case, this is a helpful learning experience for a university graduate because it clears the way for music as an avocation, or a future in other areas.

Part of working with PMP includes a constant engagement with creative conflicts. To be accountable, students need to confront these socio-musical interactions for the betterment of the song (see Video 23 on the website at www.routledge.com/9780367517632, 6:07–7:30). This video example involved final-year students and, because of this, the way I framed my advice was more forthright. I suggested that students either be accountable to the music and confront creative conflicts or don't and be happy that the music will suffer and be mediocre. Within practice-based activities, students should experience the conflict that is interrelated with creativity and engage with this on two levels: (1) a cognitive process where decisions are processed internally and (2) a practical and social process where decisions are made with others. Both of these come with their associated challenges. This was discussed with my students who suggested that "little details and tweaks" from somebody else were important in helping creative processes when an individual was "stumped" (EdD Focus group 2, October 29, 2018).

Students can experience myriad forms of conflict with varying degrees of experiential severity. All such conflicts can be useful given reflection and discussion in feedback sessions. An example of a challenging creative conflict that came up in my classes is exemplified in Video 8 (see Video 8 on the website at www.routledge.com/9780367517632, 1:18-5:00). I have mentioned this communication breakdown before but the important point for this section of the book is that, through mentorship and feedback discussions, students could reflect on ways in which they can improve. I discussed this conflict with the two students involved:

THOM: I kind of just said it without thinking, that's why it came across as just abrupt.

RESEARCHER: And what have you learnt about that?

THOM: Well, that's not very helpful.

RESEARCHER: Do you think that your persona or your vibe in the studio affects others?

THOM: Yeah. I'd say so, I mean a lot of people would know me as like saying abrupt things so I could probably like put 'em in [check] and not fight.

Elliot's comments regarding this confrontation were quite different:

ELLIOT: I think I handled that quite well … 'cause … I think almost if I went back and did it now or if I was in a different mood, that I could've quite easily … just said, "ooh, get fucked".

RESEARCHER: [Laughs]

(Thom; Elliot EdD interview 3, July 9–10, 2018)

There is a skill in navigating such conflicts, and developing this skill is vital for a career in music production. As an example of mentoring students through these issues, I introduce sayings that they can use as one method for navigating conflicts. For example, I watched an Aerosmith video (Torgoff, 1990) where Steve Tyler and Joey Kramer were engaged in an intense debate regarding a drum recording for the album *Pump* (Aerosmith, 1989). In this video, a line Tyler used to get Kramer to have another go at the part (when he was threatening to walk out) was, "hell, it's not wrong, it's just not right yet!" What a great line this is in its deflection of the connotation of blame. I tell my students about this and other sayings and these become useful to them during the creative conflicts they may encounter. This kind of tacit ability can inspire students to not run away from creative conflicts but confront them in suitable ways and increases their accountability. In my doctoral study, student improvement in this area is exemplified by the fact that, by the end of the year, the students were resembling professional practitioners in many ways: "there was so much they did today that was really just a recording session" (EdD Journal entry, Trimester 2, Tutorial 10, September 19, 2018).

I would be lying to you if I was to suggest that facilitating education and mentorships in this way is a simple process. It is massively time consuming and institutions do not allocate appropriate workload hours for this style of education. It is also different every year due to the unique personalities students bring to the classroom. Some students come with massive egos that need to be carefully deconstructed. Sometimes the educator needs to hold their own ego in check; often I suggest musical ideas to a group and the students look at me with funny looks before saying "what? … you're kidding right?! … sorry not going to happen". Educators need to practise what they preach. If we are to throw our musical ideas out to our peers (the students), there should be no ramifications if they come back with a "no!" That's how music is made. One of the most common mentorship statements I say to students to increase their accountability, control their egos and minimise emotional scarring is "it's not personal, it's just for the betterment of the music".

Educators also need to be held accountable to the team. PMPP requires an intuitive, instinctive and adaptable educator who is constantly acquiring knowledge regarding the practice of PMP but is not determined to prove their worth by overwhelmingly promoting their musical attributes (and ego) in class. This kind of personality will not be a valued mentor to students. Instead, educators should share their enthusiasm and wisdom for PMP during feedback sessions so that student accountability can be nurtured through student-centred approaches. Students agree with this conception:

> Remove lecturers who have been saying the same thing for the past 10 years, or are very stuck in their own glorious past, but don't share the meaningful side of their past, just the impressive past – like I want to know why that Grammy-award winning album you did was successful.… Share the enthusiasm of what you did and relate that meaningfully to the students.
>
> (International student survey response, May, 2021)

MAKING DECISIONS

Encouraging students to make decisions can be a very frustrating task. As discussed earlier, students will often be apprehensive about making decisions because they are scared of offending their peers. It takes time in these scenarios to develop student confidence; they must be drip fed in over the course of a student's university study. However, there is one technique that I can recommend to you as was noted in my journal notes from my doctoral studies:

> I am trying to get the students to learn to make decisions more effectively by taking emotion out of the decision-making process. They often seem afraid to make decisions that may hurt other people's feelings, but music production decisions are for the betterment of the music, they are not personal. So today I moved the decision makers into another room. This meant the artists (whom the decisions were being made about) were not present. Students suggested this made decision making easier, and subsequently I suggested to students that this physical activity needed to be contextualised in their minds when making decisions. "Simply remove the emotional aspect and make an informed decision" I told them.
>
> (EdD Journal entry, Trimester 2, Tutorial 7, August 22, 2018)

This process can be viewed in Video 28. It is a long video example but it documents the students' prolonged decision-making process and my instruction to cognitively relocate oneself to a different space when making decisions. Any mentorship advice educators can offer to the group that promotes the students making decisions more readily will, in turn, promote student social growth and empower accountability in their actions more readily (see Video 28 on the website at www.routledge.com/9780367517632).

TRUSTING YOUR INSTINCTS AND FINDING WHAT WORKS FOR YOU

One of the greatest differences between aspiring professionals and successful practitioners comes down to self-confidence. Professionals can walk into any musical context and make things happen. I spoke to Jan-Olof Gullö about this:

> Denniz PoP was the mentor of Max Martin and without him we would not have even seen Max Martin. What [Martin] said about Denniz PoP was that he was the kind of guy that, when you worked with him, you had the feeling it was going to be alright. You will have to work like hell … but we are going to make it.… Young people of today have a lot of things they are not sure about [and these things] are occupying their minds. Max Martin learned all of this from Denniz PoP [and then] worked out what worked for him. He works on one project at a time, completes it then moves on. However, he is so humble and a nice guy,

> after talking to him for two minutes you want to work with him.... He is polite, but direct [assertive]; there is no bullshit.... Success is achieved in many different ways, but it has to be actioned through a process that works for you.
>
> (Jan-Olof Gullö interview, April-May, 2021)

Whilst working on music autonomously and in classes, students should be enticed to discover the ways that work for them: the people, the places the technological spaces that help them make great music. With this knowledge on board, students gain confidence in their ability, they learn to trust their instincts and this empowers them to embody the persona of professionals. Mentorships can be useful to unpack students' personas as unique recognisable qualities; the idiosyncrasies that clients will associate with working with them. Students can then be encouraged to become more accountable towards the refinement of these qualities so as to put their best foot forward and siphon out that which doesn't work, and this clears the way for students to discover what *does* work for them. This discovery process will require students to engage with their practice, make decisions, feel the impact of mistakes and adjust future courses of action. This is tacit learning working in many socio-musical ways; even when students don't think learning is occurring, they must trust that it is.

FEAR, SOCIAL ISSUES, STUDENTS GETTING TO KNOW THEMSELVES AND STUDENT CARE

Students who engage with PMP often do not become accountable because of fear of what others may think. Regarding creative practices like music, May (1975) writes that the individual who creates has "doubts as you and I have ours, yet has the courage to move ahead in spite of these doubts" (p. 21). McNiff (1998) expresses similar notions whilst writing positively regarding making mistakes:

> There is also a beneficial sense of humility that accompanies the making of mistakes. They remind us of our frailties and they deepen appreciation for more enlightened moments. Imagine what you would be like without your mistakes. What kind of person would you be if your errors were never pointed out to you?
>
> (p. 43)

When fear is combined with an attachment to ego, the result can negatively impact music-making processes. Bayles and Orland (1994) suggest a reason for this: "while others' reactions need not cause problems for the artist, they usually do" (p. 37). For students to become accountable to themselves and engage with PMP, the students' fear of other people's opinions often needs to be addressed. To combat this issue in class, students should be instructed to "leave their ego at the door" and, with that, the need to be correct. To become more accountable, students should be empowered to face their fears and engage in unfamiliar socio-musical circumstances.

One of the students from my doctoral research (Georgio) suffers from anxiety and he discussed his ongoing battle with fear, communication and psychological interaction in the tutorials:

GEORGIO: I think I've gotten better at communication and psychological skills... I think sometimes I take a step forward and then two steps back.
RESEARCHER: Why do you think that?
GEORGIO: I reckon it's got a lot to do with my personality type and stuff that I struggle with, you know, on a daily basis.... It might stump me and instead of pushing past it, I might stop and sort of freeze which is a bit of a problem.

(EdD interview 2, April 24, 2018)

Many adolescents suffer from generalised anxiety. Harper, Marks and Nelson (2002) suggest that "generalized anxiety disorder is also common in adolescents, and excessive anxiety or worry especially about the quality of their performance or competence at school, is the hallmark" (p. 38). Student health and wellbeing needs to be carefully considered and mentorships provide the perfect role for it to be monitored regularly. Georgio, suggested that, through mentorship activities with myself, he was confronted with some realisations regarding music production professional practice and how his social traits influence this:

> Well, it's probably a combination of my personal life ... and where I am, as in emotional growth. [My] confidence has exponentially risen over the whole year but it was mainly just one day you said to me, "ah, just let go of it and just do". You didn't say it like that, but it was more like, you know, like, "fuck ... why ... why hold back? It's the only chance you're gonna get in this sort of thing". That was a turning point, yeah.
>
> (Georgio, EdD interview 4, September 24–26, 2018)

Following one year of advanced study with me and a mentorship that spanned three years, Georgio discussed his improvement whilst watching himself on a video where he was communicating with a professional session bass player (whom he had not met before) about his part (see Video 24 on the website at www.routledge.com/9780367517632, 8:50-10:08). In this video, Georgio can be seen overcoming his anxiety by communicating his ideas effectively to an industry professional as if they were colleagues. The lexicon used in this example was very musical, Georgio beatboxed sounds and, at the end, swore whilst complimenting the musician as a sign of respect before leaving.

> Yeah, great ... I think I look more animated than I thought ... than I felt at the time. Like, I felt just chilled in doing it. It felt quite natural to talk to him.... And then there was a boost ... there was a big boost of adrenalin after it, almost like, I just sat down and was just like, "aw fuck, I really did good just then!
>
> (EdD interview 4, September 24–26, 2018)

I had numerous personal conversations with Georgio where I was able to gradually develop a professional relationship with him. These were laid back social chats in the halls, after class or sitting at the café. As a result, he felt comfortable discussing his situation and I was able to positively affirm his development to him in a casual way. Subsequently, Georgio made the decision to put his fears behind him, voice his ideas and let things come as they may. He became accountable to himself and faced his anxiety; he did this because he learnt through the mentorship that it was advantageous if he wanted to produce music as a profession. This was a powerful learning outcome. To be honest, it's one of the moments I keep going back to in my mind to remind me why I do what I do. Unfortunately, within the institution I work, there are more examples of students not facing their fears than those who do.

Student care (emotional scarring)

By working with students in the way I have described above, educators have a duty to ensure the health and wellbeing of their students. During my doctoral research I developed a mentorship with one of my students (Thom) to help him improve his communication skills.

The mentorship progressed well and finally reached a point where, in his final semester, Thom felt comfortable enough to inform me that his communication problems stem from a medical history of being diagnosed with Asperger syndrome. This disorder affects a person's ability to effectively socialise and communicate. In one of our mentorship sessions, Thom watched some video footage of his interactions including the conflict with Elliot described earlier (see Video 8 on the website at www.routledge.com/9780367517632, 1:18–5:00).

THOM: Very awkward, and I was trying too hard …
RESEARCHER: When you watch that, what can you take from that now?
THOM: Oh, I can take that I actually should tell artists, "oh, maybe you should do this" instead of bottling it up and then being like, "fuck you!".

(EdD interview 4, September 24–26, 2018)

However, through continued mentorship during and outside of class (see Video 20 on the website at www.routledge.com/9780367517632), Thom's communication practices greatly improved. This was confirmed by the other students in a focus group discussion when they watched a video from earlier classes and all realised how much Thom had improved.

THOM: Well, I was wondering what the fuck I was lookin' at the computer screen for.
RESEARCHER: Yeah, so you had your back to everyone …
THOM: Yeah …
RESEARCHER: So, what … what can you take from that?
THOM: I turn around a lot more now!

ALL: [Laughter]
RESEARCHER: Yeah. You do now, hey?
HAYLEY: You turn around. You're a changed man!
ELLIOT: Turned a new Thom!
ALL: Laughter.

(Focus Group 2: October 29, 2018)

The students' laughter affirms their support for Thom's improvement with his communication. Mentorship was an important element of Thom's educational experience that promoted his individual accountability and career aspirations of being an audio and mixing engineer. We had discussions in the hallways, at cafés, gigs, during class and in my office. As a teacher in this environment, I was responsible for supervising these sometimes-confronting social exchanges. It is important to consider the damage that such experiences can potentially have on students. The educator, if approaching PMPP in this way, has an associated duty of care to their students. They need to identify potential at-risk students and do all they can to ensure these students' health and wellbeing during the learning process.

SELF-REFLECTION, DEVELOPING LIFE SKILLS, NEW BEGINNINGS AND REALISATIONS (MUSIC IS NOT FOR ME)

The approach I am promoting in this book aspires to empower students to engage with self-reflection of their music, their creative practice and themselves as social beings. I am all about providing an enriching experience that helps our students develop their life skills. I often confront students with this in class (see Video 8 on the website at www.routledge.com/9780367517632, 02:48-03:20).

> In this class, my job is to engage you with the actuality of doing music production but also to engage you with yourselves, who you actually are, what your strengths are and what your weaknesses are as a social being. I am not here to judge you; I am just here to expose you to a bit of you. So that you can reflect and go "okay I need to do this".
>
> (B. Anthony, PMPP Video 8)

By approaching PMPP in the way I have outlined in this book, students can be encouraged to learn about their social qualities, whether they talk too much, whether they don't talk enough, whether they can be rude or whether they are just too passive. In my experience, students develop negative traits in reaction to their confidence gap (Harris, 2010) – their link between fear and personal confidence. By utilising mentorships and working with students on their accountability, educators bear the possibilities for life-changing learning to occur within the student. Oftentimes accountability, in these instances, is empowered through a student's self-belief. Harris' (2010) confidence gap ideas aligns with active participatory learning (Ciobanu, 2017) which relies "not only on the cognitive skills of learning but also on the mental and emotional ones" (p. 167). If students are developing their accountability, they will embody active participatory learning where "learning is a personal activity that belongs to the individual [and is] decisive in shaping [a student's] personality" (Ciobanu, 2017, p. 167). This is some of the greatest learning popular music students can do: learning to face their fear of who they are and to be confident with their engagement in PMP.

In my research it was clear that three of the primary participants, all of whom I had developed musical mentorships with, had learnt a tremendous amount about themselves. They had used the class experience and mentorship to do so, but it was as if they were enlightened about themselves, and where they wanted to go next in their lives.

Thom is much more comfortable communicating in the studio now and has begun to forge a career as a recording and mixing engineer. Georgio has taken positive action to confront his anxiety when making music. He has completed first class honours (under my supervision), is producing and performing original material and educating young musicians on synthesiser and bass guitar applications via face-to-face lessons and his YouTube channel. Lennon was one of the strongest songwriters of the cohort but she has decided that she wants to work in A&R. Lennon realised that she doesn't want to perform and record her music, she wants to use her talents combined with her street-sense and engage in the life of a record company executive. These are things the students worked out about themselves. Through my courses and mentorship, I aspired to educate students about PMP, but it turns out students were developing life skills as well and charting their next steps towards their futures. Perhaps this is the greatest lesson of all.

REFERENCES

Aerosmith, (1989). *Pump [CD]* [Recorded by Aerosmith]. Vancouver: Geffen.

Anthony, B. (2019). *Perspectives on learning popular music production in higher education from "both sides of the glass"* (Doctoral dissertation, Griffith University).

Bayles, D., & Orland, T. (1994). *Art and fear.* Santa Cruz: CA: The Image Continuum.

Ciobanu, N. (2017). The applicability of active-participatory methods. *International Journal of Humanities Social Sciences and Education, 4*(11). DOI: 10.20431/2349-0381.0411019

Cumming, N. (2000). *The sonic self: Musical subjectivity and signification.* Bloomington: Indiana University Press.

Harper, G., Marks, A., & Nelson, W. M. (2002). Teen depression: Overlooked and undertreated. *Patient Care, 36*(12), 37–47. Retrieved from https://link.galegroup.com/apps/doc/A94044544/ITOF?u=griffith&sid=ITOF&xid=44a016af

Harris, R. (2010). *The confidence gap: From fear to freedom.* Melbourne: Penguin.

May, R. (1975). *The courage to create* (Vol. 1). New York: Norton.

McNiff, S. (1998). *Trust the process: An artist's guide to letting go* (Vol. 1). Boston, MA: Shambhala.

Torgoff, M. (1990). *Aerosmith: The Making of Pump,* [Video]. U.S.A: CBS.

Zak, A. (2001). *The poetics of rock: Cutting tracks, making records.* Los Angeles, CA: University of California Press.

Part 4

Pedagogical timeline, graduate considerations and summary

This section of the book brings together many of the pedagogical concepts that have been broached throughout. It presents (1) a suggested timeline of delivery, (2) post-undergraduate considerations and (3) a summary of the book's recommendations. This section is designed to give the reader insight into the ways I have facilitated PMPP in the past. These recommendations may then be contextualised (if deemed appropriate) into various educational experiences.

DOI: 10.4324/9781003055099-24

20

Student progression and graduate futures

There are optimum times to engage students with the many concepts discussed in this book. The student progressions I have outlined below have been developed to work within the guidelines of my institution's programs and the cohort with which I currently work (B.Mus. popular music and creative music technology majors). I offer them to you as possible influence for your own designs.

The first timeline (see Figure 20.1) embodies my attestation that all students of PMP, no matter if they are an electronic or acoustic-based musician, should undergo some foundational training including basic sound recording, digital music theory and music programming. This should be framed within acoustic, electronic and online environments so students develop a holistic knowledge base. It is also beneficial if students can take additional courses in songwriting and musicianship, such as music theory and performance. In second year, students can learn to apply this knowledge in diverse creative practice scenarios (electronic, acoustic and online) and these activities can be nuanced and weighted in accordance with the various programs students are studying. Over a student's final years, I find there is value in allowing them the freedom to design their music-making activities towards the goals with which they aspire (student-centred learning). This affords students the opportunity to take the knowledge they have obtained in years one and two and begin shaping their practice and future aspirations as portfolio musicians and popular music producers. The key here is that, with additional educator mentorship, limitations in students' practices and aspirations can be articulated. You will also note that student autonomous work in the aforementioned figure always enables the student the opportunity to work on original material that interests them. They may also be given some specific genre or musical assessment tasked by the educator, but I believe in letting students choose some aspect of their work in all years of their study as this ensures their interests are nurtured throughout their university lifecycle.

DOI: 10.4324/9781003055099-25

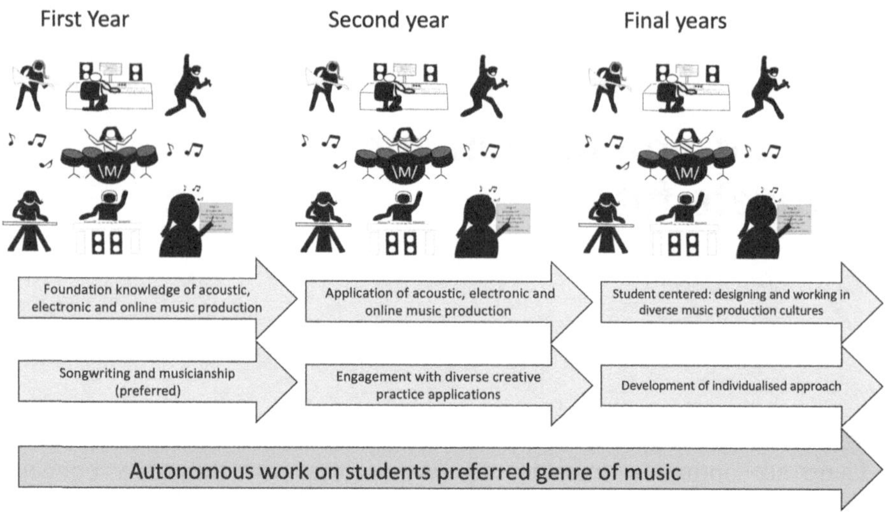

Figure 20.1 PMPP foundational knowledge, creative practice and student-centred timeline.

ADVANCED POPULAR MUSIC PRODUCTION PEDAGOGY TIMELINE

Throughout this book I have outlined my ideas regarding the learning of advanced concepts of PMP. In the second timeline (that is exemplified in Figure 20.2), I present the periods in which some of the concepts explored in this book can be actioned. Some elements, like PMP skills (songwriting, programming, musicality, etc.) and a student's aural library, are things that begin in elementary ways prior to the HE experience. These are then developed further with students from first year and beyond. Figure 20.2 also shows how the educational experience can be staggered over the three to four-year span of an undergraduate degree to offer a transitional approach regarding the more difficult concepts like student accountability, soft skills and critical thinking. The educator's role is also essential to this approach because it diversifies over time from more formal teacher-led approaches in first year (skill acquisition, for example.), to the turbocharged and student-led models in final years.

Pedagogy of this type enables all students to engage with every aspect of PMP and student-centred approaches help to generate a diverse array of practitioners. This is supported through mentorships so that students can discuss their skills as producers, identify the applications with which they are suited and consider any applications they may need to outsource in various music-making experiences.

Research led practice, the road to post-grad, masters and beyond

I spoke at length with Paul Thompson, Russ Hepworth Sawyer, Rob Toulson and Mark Marrington about the PMP Masters programs they

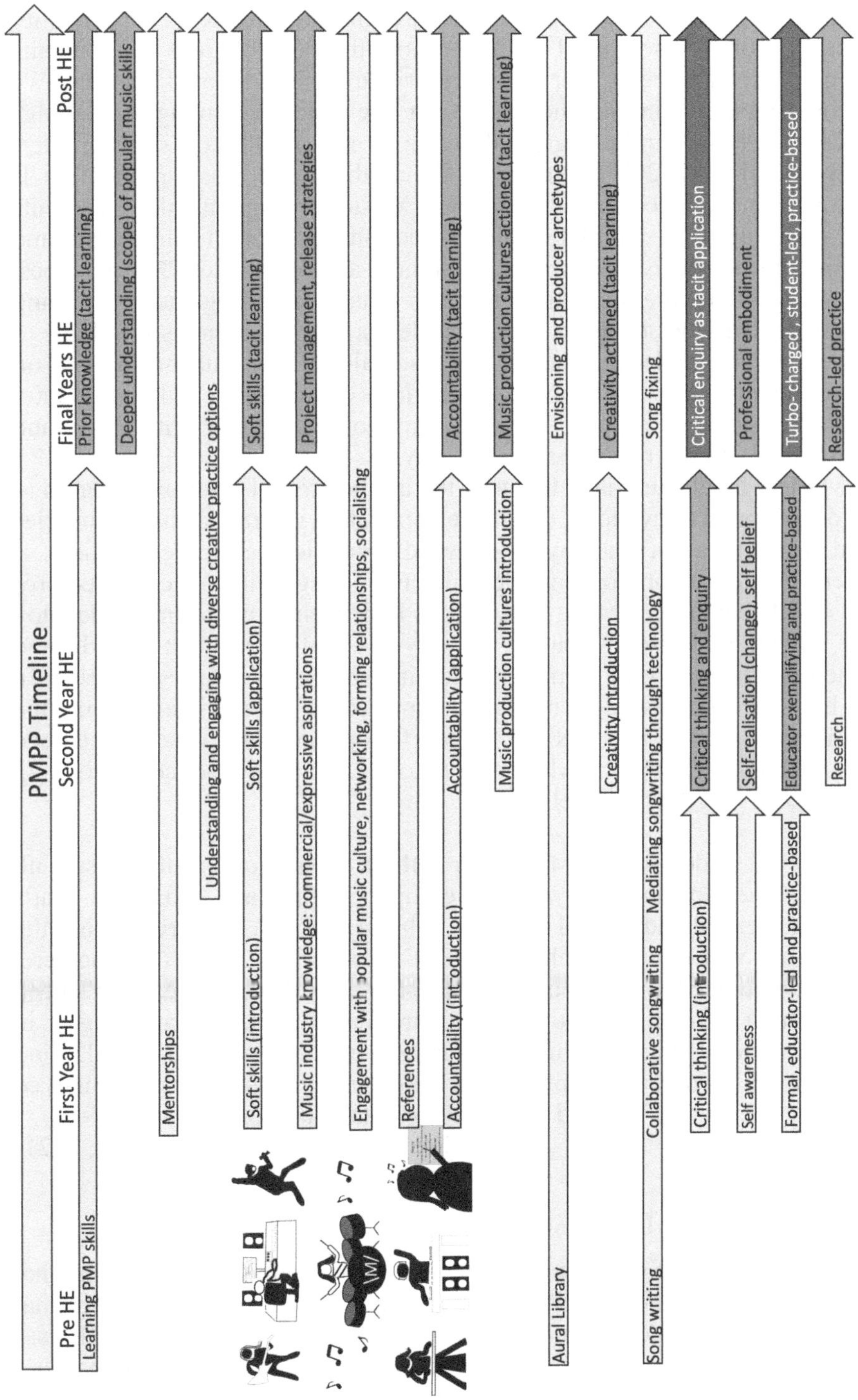

Figure 20.2 Popular music production pedagogical timeline.

have been involved with and, I must admit, I wish my institution targeted Masters programs such as these more. I often say to my students, "you need about ten good years of making music to get anywhere near knowing what the hell it is you're doing and if you are, in fact, producing high-quality music". Rob Toulson and I discussed how the PMP degrees that students complete over three or four years are primarily foundational learning. We recognised that many institutions do not like educators saying these things because it devalues the degrees they are trying to sell (Rob Toulson interview, April-May, 2021). However, the truth is that music production is learnt over extended time periods and Master's programs allow students the opportunity to complete their foundational education (undergrad) and utilise this in research-led practice that more acutely utilises PMP practices. This aligns with the previous comments I made regarding students finding methods that work for them. Research-led practice is a fantastic vehicle to develop this knowledge. This kind of study also allows students to move on from the *how to do* PMP landscapes so they can investigate the more integral socio-musical aspects of PMP – how to make music with others and why things should be done a certain way.

Masters level study also includes the application of rigorous research to inform domain knowledge, extensive application of practice to test theories and data generation that may be analysed to develop new understandings. It opens the door to one-on-one supervision meetings with an accredited professional/academic for rich discussions and further mentoring to develop the student's skillset. It paves the way for further academic study and potential teaching opportunities following extensive industry practice. In all honesty, students might not feel the immediate urge to integrate into the professional world and this is a wonderful option in this case. I discussed this with Paul Thompson who referred to research-led practice as "looking the wrong way down the telescope".

> We get students in their final year to think of a project and they just think of something and go," yeah I'm going to do this" and I go, "cool, that's out there already", and they go, "oh". And I say, "you're looking the wrong way down the telescope; you're focused on this little thing here, but you need to start looking at the stars, what's out there, what's been written about this already, and then from reading it, what are the gaps in knowledge, what are things that maybe people have overlooked? Find something that you're passionate about and go "oh, I'm really annoyed that they haven't talked about this' and take that on as a project".
>
> (Paul Thompson interview, April-May, 2021)

ROADBLOCKS TO LEARNING

The participants of my research felt there were roadblocks to learning that could be addressed in PMPP contexts. These included bad social habits that students had to unlearn, unaccountable students in the tutorials and class numbers that were too large. However, to facilitate diverse music production cultures within the tutorial cohort, often up to six band members were required, and many technological designs that employed a large amount of equipment required three engineer personnel to operate and a group to set up and pull down. In these situations, large numbers were necessary. Yet, some students' survey responses suggested class sizes under 10 were desirable. Perhaps this is understandable since students crave one-to-one critique of their process and music. However, with the unpredictability of student engagement and some students not even showing up or being sick, it is regularly difficult to facilitate a suitably diverse array of music production cultures without at least 10–15 students enrolled in a tutorial.

One must also consider that, when working on electronic music production or mixing, small class numbers are important to replicate the profession. In these instances, small groups of students (two to three) working concurrently in multiple rooms (small post-production rooms if they are available) or even on a laptop with a headphone splitter in a suitable location are viable options. Added to this discussion is the current corporate structure and commercialised nature of the university sector in many countries around the world. Lecturer-to-student ratios (financial considerations) are always being scrutinised so small numbers in classes are often questioned. There is much to contemplate here on a case-by-case basis where a PMP course focus (electronic music, for example) should dictate suitable class numbers.

Students from my international survey also responded negatively to programs that did not allow students to produce original material, did not allow them to work on the genre of music they preferred and maintained a fascination with large format recording studios and the associated dominance of production processes from past eras. Students also did not value courses that focussed on technological application:

> I think that many universities are too focused on the technological side of music production, and not on the emotional side of it, which is arguably more important to creating a great piece of music.
>
> (International student survey response, May, 2021)

Students also suggested that many educators from around the globe were not empowering and nurturing human beings and they dominated students' practices with their own opinions and discussions on their own works.

> Keep the creativity in mind throughout the whole thing. Don't try to push specific agendas (the teacher's) on people unless they ask for it. Music is so broad these days and such a tough industry that a lot of people aren't going to make it but they can still have fun doing what they love and producing.
>
> (International student survey response, May, 2021)

Students felt that unqualified educators also impacted the pedagogical experience in a negative way:

> In my experience I have found that there is a struggle to find qualified professionals in the industry that also meet post-secondary educational

> requirements [a masters or PhD]. This means that students miss out on some valuable current industry information.
>
> (International student survey response, May, 2021)

Most of the roadblocks to learning that I have encountered (as listed above) come from any influences that disrupt the educational experience from a real-world context. Professionals ensure the socio-musical characteristics of a music production culture are suitable for the session. The right amount of people are in the room, bad attitudes are addressed and removed and creative practice is focussed on modern applications that suit the musical intentions of the artist. Educators need to do all possible to replicate this practice in the classroom or, alternatively, highlight the times when a class is not doing this so learning ensues for all.

LIFE-LONG LEARNING

The learning of PMP, as such, never ends; there is no final destination. Learning continues until we stop making or listening to music. Therefore, our students benefit if they are equipped with the knowledge of:

ALEXIS: ... how to learn in a variety of ways, because how you learn changes over time and is reliant on the resources you have available to you. You need to be collaborating, working things out and using all these things that have been dubbed as soft skills … to drive life-long learning. But before you do life-long learning you need life-long questioning.

BRENDAN: Is that where this ties into critical thinking?

ALEXIS: Yeah exactly.

(Alexis Kallio interview, April-May, 2021)

Rob Toulson discussed the importance of autonomy and life-long learning:

> You know the whole concept of autonomy is still right! I think it takes a good ten years to become really good at something, you know, until you're good enough for someone to pay you to do it. Uni is only a small part of that … it's like 30% of a learning cycle. I try to make it clear [to my students] that uni is just the start. If I am honest I think it takes 20 years! When you're 40 you've then got the knowledge, the experience, the wisdom, you've got confidence, some contacts, you've seen some failures and success. That's how you learn it, but it never ends.
>
> (Rob Toulson interview, April-May, 2021)

Our students need to appreciate that PMP, as a form of agency, is not realised through the manifestation of a high-quality piece of music, it is learnt over time. Life-long learning can be unpacked within ongoing discussions with students, and this is one of the main reasons I continue to mentor graduates of all kinds well into their professional careers. I have online chats with touring musicians, producers who are setting up their own home facility, songwriters, school teachers, graduates now working in record companies, in police forensics and music therapy. This extended pedagogical approach (life-long learning and mentorship) can be used to encourage graduates to always be open to new ways of making and utilising their knowledge of music. An acceptance of life-long learning prepares our graduates for the realities associated with an innovative professional music practitioner of the future.

GRADUATE FUTURES

I am sure I am not alone when I say there is no shortage of great music on this planet, it's just that we haven't heard most of it! Unfortunately, the facilitation of a decent education in PMP seems almost pointless if our graduates are not adaptable and considering innovative futures.

> Your best students are the ones who see opportunity.... We teach Peter Drucker for example.... We get them to think about, how can they create something for themselves? How can they disrupt what's going on in front of them? How can they change things?
>
> (Russ Hepworth-Sawyer interview, April-May, 2021)

Prince Charles Alexander and I discussed what graduates of the future should look like:

> I don't believe in the master of one thing. I think a jack of all trades is the way to go. I think if you are top-liner and you can record your own vocals and then do your own doubles and add some nice sweet reverb and delays on your vocals that pop off at the right places. You've just increased your value.
>
> (Prince Charles Alexander interview, April-May, 2021)

Both Russ and Prince Charles Alexander are promoting graduates that are multi-skilled and adaptable. I think one great example of a PMP graduate (in the pop world) is Charlie Puth. I watched a video (Rolling stone, 2017) the other day of him describing his practice. He can write, produce, record, program, play, sing and perform – multi-skilled. Another great example of a multi skilled graduate of PMPP from a completely different perspective is the proofreader of this book: Sam Vallen. Sam has a Ph.D. and is the co-founder of progressive rock band Caligula's Horse who are signed to Inside Out Music (Dream Theater, Devin Townsend). Sam writes, produces his own and other artists' works, programs, goes on world tours and is an educator at higher education – multi skilled. Both Charlie and Sam (in their own ways) have forged a future in the music industry by understanding and integrating into the system while, at the same time, making themselves adaptable enough to work in many ways.

In my opinion, educators must confront their students with the financial realities of the current system and empower them to become multidisciplinary. That way, if they choose PMP as a career, they can make a living in various trajectories. Educators must also empower students to consider new innovative alternatives that will disrupt the system and revolutionise the popular music industry as an equitable, inclusive, diverse and commercially viable option for all.

THE VALUE OF PMP INTO THE FUTURE

We must not lose focus of the value of PMP as a resource for our students' futures. In reaction to recent industry changes, I have noticed that some institutions have adopted an approach that seems to devalue the practice of PMP in favour of performance-based courses. Rationales behind this promote the idea that songwriting and touring are guaranteed forms of income for popular music graduates. While this may be true, and there is no denying the integral importance of songwriting and performance education, it is my opinion that education in these two areas alone may potentially locate the popular musician of the future within endless tours where they will barely make enough money to survive. In my discussions with Sam Vallen, he suggested that "very few small and medium-sized artists actually generate much income from touring the world nowadays – it took us (Caligula's Horse) three European tours to turn our first real profit, for example" (Sam Vallen, personal correspondence, January, 2022).

Yes, streaming platforms have minimised mechanical sales, but recorded works now have increased value as promotional vehicles. Streams, views, etc., now equate to a monolithic promotion and marketing resource, replacing radio airplay as the primary way an artist's brand is presented to the public. Without high-quality recorded works, aspiring professionals will not be able to service the requirements of Internet-based promotional mediums. Therefore, all popular musicians should be able to contribute to the production of their music in some meaningful way.

REFERENCE

Rolling Stone, (2017). See Charlie Puth break down emotional hit song, Attention. YouTube video, Retrieved from: https://www.youtube.com/watch?v=IU8BEMi8UyM

21

Pedagogical summary and challenges

Anyone with a suitable computer is now capable of engaging with music programming and recording. As such, all popular musicians (including songwriters and artists) have the potential to be considered popular music producers. I believe it is our duty as educators to ensure all students engage in this practice as much as possible so as to better prepare them for their futures. Post-graduation, some students may go on to use music as an avocation and if this is the case, effective pedagogies should have opened their eyes to the endless possibilities that their music making may include. Many students will seek integration into the music industry and, therefore, effective PMP educational frameworks must also offer a modern representation of the practice of PMP and aspire "towards greater relevance and career preparation" (Stewart, 2010, p. 244). This is crucial so that graduates are able to "navigate the working world and successfully manage the career building process" (Bridgestock, 2009, p. 35).

My research engaged students with many of these modern perceptions and found that effective PMPP developed someone's understanding of the greater *practice* of PMP. Students also developed tacit knowledge through an engagement with some key pedagogical schemas that have been outlined in this book. I have also supported these findings or presented differing perspectives through a multitude of statements from esteemed educators and students from around the globe. These schemas highlighted the usefulness of music production cultures, the importance of an intuitive and instinctive educator who was an industry professional and research enthusiast, the necessity of bridging the tutorial experience to the professional world so that students experience the socio-musical realities of PMP and the many ways in which students learnt PMP. Effective pedagogy emphasised student-centred approaches that enabled students to learn about the varied musical landscapes within which they can work and create, how to create different types of music that conforms to these landscapes and how to apply skills that are explicitly informed by the music production cultures in which they operate. Students should also leave university with a modern understanding of the music industry and potential release strategies (developed through project management activities) so they can integrate into the current system, or else innovate new pathways. If students

DOI: 10.4324/9781003055099-26

participated in pedagogy that was shaped by these, it was found that they were given the opportunity to engage with their strengths, improve their weaknesses, develop confidence and prepare for the social, cultural and musical realities of modern music making.

CREATING COLLABORATIVE MUSIC PRODUCTION CULTURES IN HIGHER EDUCATION

This book has identified some approaches that HE could incorporate to create and facilitate effective PMPP. These are grouped below under domain cluster headings from my doctoral research. My tutorials in the past have achieved some of these schemas, however, it was the journey that students undertook which influenced student reflections and discussions, and these were the most important catalysts for student growth.

Schema: The impact of diverse music production cultures

- Effective contexts for learning popular music production engage students in many diverse music production cultures. These include acoustic, electronic and online collaborative processes.
- The ideological perspective of the pedagogy comes from both sides of the glass.
- Student cohorts that possess prior learning in a diverse range of popular music skills (musicians, producers, engineers, programmers and songwriters) enable music production cultures to occur.
- Music production cultures engage students with diverse socio-musical contexts from which communicative and psychological skills can be developed tacitly.
- Students' development of music production skills are explicitly informed by the overarching influence of the music production cultures in which they operate.

By placing students into diverse music production cultures, educators do all possible to replicate the embodied learning that I experienced as an assistant engineer and professional practitioner. In the industry, one day I could be working on Hard Rock, the next could be Pop, Rap, Surf Rock or Film music and my role would shift from producer, songwriter, programmer, session musician, singer, third-party mixer or a combination of any of these. I also developed an appreciation for the collaborative efforts of all the people involved in the music-making process. However, I have little doubt that it was my ability to work in diverse ways that helped achieve my longevity as a music-making professional. PMPP should mirror these experiences but also ensure the inclusion of student-centred contexts – working with the music to which student's feel connected. By facilitating a broad scope of PMP engagements within diverse music production cultures, educators provide students with an experience that can inform their future music-making possibilities.

Music production cultures provide the opportunity for advanced PMP learning outcomes to occur in HE. They represent a concept that can align educational approaches across the globe. Uniformity in this regard will ensure all programs are moving past foundational education in sound recording or music programming and, alternatively, develop in students a nuanced understanding of PMP as a greater practice. The use of music production cultures most importantly considers the individual socio-musical and cultural characteristics of all types of music and the people that engage with them so that education is equitable and inclusive for all.

Schema: Student accountability and engagement with the professional world

- The tutorial experience and assessment items utilise diverse recording environments. These include home studios, traditional recording studios, MIDI-labs and post-production suites.
- Access to a collection of historical, electronic and modern music production technologies is optimum.
- The stigma that is associated with the university building is considered so that the *place* that learning happens emits an atmosphere that inspires music making.
- Student accountability is targeted from two perspectives: (1) to themselves, and (2) to the group.
- Students are confronted with the realities of being either hobbyists or professionals and an emphasis is placed on the mantra "how you practice is how you play".
- Musical activities and assessment items include covers and student originals from a diverse range of genres.
- Time limits place students under pressure which bridges the pedagogical experience to the professional world, promotes student accountability and highlights student inadequacies which can be learnt from and improved upon.
- Students work autonomously outside the classroom environment by organising, discussing, networking, rehearsing and liaising as if operating in the professional world.
- Professional practitioners (session musicians, producers) are brought into the classroom to engage in the practice-based activities.
- Students should be challenged to consider the commercial or expressive function of their musical works and design release strategies associated with these.

I think one of the biggest takeaways from the writing of this book is that a student's engagement with HE is ultimately about providing an enriching experience. Within this context, students should critically consider who they are as musicians and what they may, or may not, want to achieve. Additionally, music educators can help students become valuable and respected people in society. As such, it is important that students are made to feel accountable for their own development as learners. This then transfers to their life in general, becoming accountable to their ever-evolving development as people. Their social skills, their roles in society and how they either integrate, diffuse or confront societal situations are shaped by what we do. Therefore, in contrast to directing them, educators should work towards the encouragement of students to take control of their HE experience and empower them to consider their impact on music making. This may include social interactions that mirror their engagement with society in general, and students can learn how they may improve in these areas. We are ultimately giving students a "trial" experience of modern life, but one where it is a safe place to make mistakes because this promotes learning. By helping students confront these difficult territories, educators are preparing graduates who are hopefully open to change (even within themselves) and looking towards futures where they consider what is right and wrong in the world, and how to manifest change where it is necessary.

Schema: The educator's role

- The educator is an experienced industry professional and has, or is in the process of, conducting relevant postgraduate research.
- The educator's role is instinctive and adapts to student engagement utilising formal instruction, becoming a music production sounding board, facilitating turbocharged learning and becoming a music production colleague.
- The educator facilitates reality checks to students when their engagement is poor, almost epitomising a workplace boss at times.
- Transference of ownership of the tutorial experience to the students is an end goal of the pedagogical approach.
- The educator develops professional relationships and musical mentorships with the students which nurtures student self-development.
- The educator uses a lexicon that students relate to, nurtures the relationships discussed above and breaks down the hierarchal dominance of master/apprentice pedagogical frameworks.
- The educator uses monologues to outline the tutorial activities and bridge the tutorial experience to professional practice by discussing relevant scenarios.
- The educator can utilise mentorships with students to have discussions and listen to music. This will help to update the educator's knowledge of music production applications that are deemed current in the industry.
- There should be a sense of equality to the educator/student dynamic.

The role of the educator is the one area where I feel the most frustration and where I believe institutions have considerable work to do. I also feel this is justified through many of the negative comments made by students about their educators as collected by my survey. The pedagogy discussed throughout this book requires a well-balanced educator. In my opinion, educators need to develop their skillset so it includes a balance of professional experience and an engagement with pedagogical research and/ or research-led practice. An educator with limited professional experience who relies primarily on scholarly reading is in danger of approaching PMP education from an unrealistic perspective. However, I have also seen many educators with astounding professional careers but who, when I've heard them talk, have made me think "man this person really sucks at being an educator". Mostly, this has to do with their idealistic viewpoint and inability to refrain from self-glorification to focus, instead, on students' needs.

If I see a pedagogical PowerPoint with an educator's face and the records they worked on, automatically I question the pedagogical standpoint of that person. A person's credit list does not justify their place and suitability as a PMP educator. A person's willingness to help students discover who they are and what they want to do as musicians is a key aspect to facilitating an education in PMP. Students connect more deeply with authentic communications from a genuine and empathetic human being. Someone who has taken the time to develop a relationship with them and get to know them, someone they feel connected to, someone of equal standing. If that person is then recognised to have an eminent career, this only enhances learning.

As experienced professionals, I believe it is our duty to draw on our experience and empower students to unpack their vision for an innovative future. As educators, I believe we must locate ourselves at a viewpoint that values professional experience alongside pedagogical rigour whilst ourselves being empathetic, authentic human beings. I have not discovered a better combination.

Schema: Learning occurs in many ways

- Lectures, tutorials (labs), student autonomous work and assessment should work together to develop student sensemaking and impart knowledge through formal, practice-based and research-led frameworks.
- Lectures should be engaging, inclusive of student interaction, and, if possible, be fun.
- Listening sessions and critical thinking are integral elements that should be undertaken throughout the educational experience.
- Students are informed that learning music production is facilitated through discussing, planning, doing, and reflecting on diverse music production cultures.
- Some aspects of the social landscape of greater university life (bullying and intimidation) can become toxic in the tutorial space.
- Other aspects of the social landscape of greater university life (band members that have parted ways) can be useful social dynamics that help the students understand the realities of music making.
- The socio-musical culture that is synonymous with popular music should be nurtured within the HE setting.
- Mobile phones and social media should be discouraged unless they are useful to the music-making activities (tuner apps, YouTube reference material, Spotify, organised chat rooms, etc.).
- There is merit in educating students about historical approaches to music production because these can be used as methods of influence on music production creative practice.

- PMPP can be enhanced with the application of a *Systems Centered Learning* (McIntyre et al., 2018) inspired approach. See Anthony's adaptation in Figure 16.1.
- Music production pedagogy uses a balance of practice-based activities that (1) produce release-ready material and (2) are learning journeys where the standard of the completed product is not important.
- The cognitive processes of music production (aural library, the producer's vision, producer archetypes) help students design music production cultures and select creative practice options.
- Students learn the skills of music production by developing a deeper understanding of the skill.
- Prolonged engagement with communication, critical listening, musicality and engagement with emotional transference develops student ability in this area;
- Soft skill development should be targeted over the entirety of a student's study because these skills take time to develop.
- Students traverse the subjectivity gap by presenting their ideas, listening to feedback and working towards aesthetic goals.
- The class environment represents a safe place where mistakes are okay and can be learnt from within feedback sessions with peers.
- A collaborative context encourages students to experience and reflect together as part of a team – a collaborative hub.
- Elements of learning music production are tacit or implicit, and learning often occurs when there is no awareness of learning happening.
- Music is a common language that can be used by all to promote a pedagogical environment of musicians as compared to an educator and their students.
- The lexicon used in the tutorials aligns with the music production culture that the group develops and includes humour and swearing if suitable.
- Students express emotion, confront each other on problematic issues, face creative conflicts, and shape their social interaction to the people with which they are engaging.
- Student learning includes a level of self-reflection – where their personality sits within the social spectrum.
- Introspective contemplation aids with students' development of life skills and by developing these further students ready themselves for their future as music practitioners.
- Feedback sessions during tutorial activity allow process learning through discussion and reflection by students.
- The tutorial is a profession-based music-making experience, a class and something in between these two at times.
- Tutorials are two-hour learning tutorials or four-hour recording sessions (or similar).
- Class numbers of 10–15 students provide the necessary resources for the overarching context of both sides of the glass and diverse music production cultures to occur.
- Courses that focus on electronic music production may find smaller class numbers more suitable.

- Students develop an understanding of the holistic practice of popular music production in a host of diverse contexts.
- Students learn about the various musical landscapes they can work within, the different ways they can make music to conform to that landscape and the application of skills which are explicitly informed by the music production cultures they create.
- For some students PMP will become an avocation not a career, this aspiration needs to be identified as valid and students nurtured towards suitable goals.
- Learning popular music production is a lifelong process; it never ends.

The number of dot points in this this last cluster emphasises the necessity for a holistic approach to PMPP that targets learning in many ways and on many levels. I can assure you that I don't think what I have written in this book epitomises all the ways we need to consider PMPP. I anxiously await more books on this subject so that we can all learn from each other. I think that ego and attachment to the ways of one's past present constant problems. We must not think that we individually hold all the answers. Rather, we must value the experiences and opinions of others. I trust this book is reflective of this approach.

Within the PMPP discourse, this book targets education in the advanced concepts of PMP through a student's engagement with music production cultures both in their classwork and in autonomous sessions. By doing so, learning takes place on many levels, but, importantly, it occurs as a form of tacit skill development. This often happens without students realising it is occurring and it culminates in a term professionals use – "experience". To achieve this, students need to intuitively apply critical enquiry, ask bigger questions that contemplate why something should be done and for what purpose, as opposed to how and this helps students become innovative music makers. These abilities are nurtured through an understanding of advanced PMP concepts like creativity, envisioning, project management and the evolving nature of the socio-musical contexts with which they work.

Courses in PMP are also part of a bigger system: a higher education institution. The institution's cultural influence must not be underestimated or neglected. Popular music culture should be ever-present in the experiences of the degree program. This should occur both inside and outside of the classroom and be represented in the lexicon of educators and students. I believe it is easy to forget that most of the PMP learning occurs in autonomous sessions and through networking and communication with others. Conversations between students and educators in the halls, at gigs or cafes should be prolific. Perhaps they would be similar to: "hey have a listen to this!" ..."Wow that's cool. Where did you hear of that?" ... "Yeah it's Taylor's new music, I went to their gig last night" ... "Sick! The lyrics and the groove are dope! You should go talk them and see if they can write a tune with you!" Popular music making has been learnt and made this way for decades. Students, like us before them, need to understand this, and then given the freedom to learn and apply their craft in the ways that work for them.

The learning of PMP is enhanced within group collaborative contexts but it is also specific for the individual. Most importantly, if you follow the approach suggested in this book, I trust you will notice that effective learning culminates when the educator and students learn together whilst making, and talking about, music. In this space there is no master/apprentice model and there is no re-imaging of a student as master model. There are just peers working and learning together. This is the end goal of the PMP HE experience: it should eventually facilitate transference of musical information between mutually invested peers.

EQUITY, DIVERSITY AND INCLUSIVENESS (EDI) IN PMPP

When I started the interview process for this book, I sent all participants a question about the greatest challenges we face as educators in this field. The one topic that raised its head time and again was the problematic history that PMP as a profession and PMPP as an educational experience had regarding matters of equity, diversity and inclusiveness (EDI). This was often to do with considerations of gender, race and social status. This book does not have the scope to investigate these topics in a rigorous fashion and, through my discussions with Alexis Kallio, it became clear that, instead of writing a separate chapter on this topic, the message should be interwoven throughout the entire text (Alexis Kallio, interview April-May, 2021). While it has been my intention to do this, there is some important content regarding EDI issues that came from the interviews, and I feel obliged to pass these on so that it is firmly understood that change is needed.

First, I think it is important (at this point) to identify a few uncomfortable truths about the industry we work in. Amandine Pras has done some exceptional research surrounding gender inequality in the music production industry (Brooks et al., 2021). The study collected data from 387 producers, recording engineers and studio assistants from 46 different countries. When compared with the results of a study surrounding women's experiences of micro-aggressions within STEM academia (science, technology, engineering and mathematics) the results showed "that the recording studio workplace scores 33% worse on the silencing and marginalization of women, 33% worse on gender-related workplace micro aggressions, and 24% worse on sexual objectification" (p. 238). It is also no secret that the number of women or non-gender specific people that are working in music production roles are not as visible as those that identify as male. This is most unfortunate. I was trained at Rhino by two people: Paula Jones and David Nicholas. Paula was, and still is a wealth of information and a single embodiment, among many, of the fact that gender plays no role whatsoever in audio engineering or production ability.

When I spoke with Amandine Pras, we openly discussed many personal reflections of inequality in the music industry, the dominance of males in high profile positions in the recording industry and the in-ethical treatment of people in non-Western countries. In particular, Amandine described a number of experiences that highlighted the inability for people who did not identify as *straight white men* to achieve high-level success in the music industry as it operates currently (Amandine Pras interview, April-May, 2021).

I also spoke with Kaley Eaton. Kaley's research illuminates the potential negative influence the current state of play in the industry can have on educational institutions. She suggests that

> the erasure from the academic canon of people of color, women, and artists of all stripes who embraced popular music, as well as the problematic power dynamics at play in the larger music industry, speaks to the vortex of power in this field that has favored white men, economic privilege, and academic prestige.
>
> (Eaton, 2021)

I had a wonderful discussion with Kaley about many things music production-related, but she suggested that many of these issues – gender, socio-economic bias, etc. – are inter-connected, so solutions are complex.

Sam Bennett also attested to the challenge we face in this area:

SAMANTHA: I mean the fact is. The audio industry has a whiteness problem; it has a patriarchy problem. It tends to elevate the work of cisgender white men.

BRENDAN: What would you say is our way through?

SAMANTHA: Well, the way through is to normalise it, to celebrate the brilliance of engineers and producers who are not white men.

(Samantha Bennett interview, April-May, 2021)

Samantha went on to discuss the micro-aggressions she regularly faces, such as people incorrectly associating the work that she is doing with her male counterparts. Nyssim Lefford has had similar experiences:

> What was that Joni Mitchell quote? "Whatever man happened to be standing in the studio that day took credit for my genius". When an idea is put forward by a woman, it is either ignored or not perceived as originating with her. (This is one reason why there are so few female producers). And, the older women get, the more invisible they become. Students see this [happening to middle-aged female professors], which both undermines the authority of women at the peak of their experience and expertise, and also, I suspect, can subliminally encourage some young women to conform to non-threatening, cute ideals.
>
> (Nyssim Lefford questionnaire reply, April-May, 2021)

Some inspiring educators are actioning change in small areas, like Annelise Noronha, who has facilitated music production courses for transgender folk (Annelise Noronha interview, April-May, 2021). However, many participants that I discussed this with agreed that the problem stems from a greater issue: outdated and unethical societal mindsets.

The suburb I grew up in is encapsulated within a low socio-economic area. Given the amount of Hip Hop that is now being composed in those suburbs, I always look for applicants to my institution's courses from those areas. Sadly, they rarely come. I often wonder if students who produce Rap, Hip Hop or similar styles just don't see value in the university experience. Perhaps the Western classical music stigma that is synonymous with a *conservatoire,* or the cost of university and related software or equipment, scares lower socio-economic students away. To combat this, Kaley Eaton suggested some very interesting universal solutions that are informed from her experience in elementary education. Things like using open-source software that anyone can access so the gatekeepers who provide access to programs that cost substantial amounts have less influence on the education system (Kaley Eaton interview, April-May, 2021).

Amandine Pras also discussed how she currently removes inappropriate behaviour from the classroom that has been accepted in the industry in the past.

> The thing that I've been reflecting upon is that there's something about the culture of the recording studio worldwide. I've seen it; it's really putting the musicians and also the artists and technicians in a kind of opposite type of social relationship [Musicians are in power positions over engineer personnel]. It's almost like the musicians are all about themselves and it's all about their feelings and they can react extremely spontaneously, almost like very young kids, demanding or whatever. Even if they're super respectful people, they get in the studio … and they're kind of egocentric about the way they can behave. And we used to teach our audio engineers to have this ability, to be distant from their own feelings and be accommodating to the musicians. So, I had professors who mentioned that to me, so it was kind of like "you are there for a reason … you can handle it".
>
> (Amandine Pras interview, April-May, 2021)

Amandine went on to suggest that, in the past, these activities were reproduced unconsciously in her classroom; they occurred when she did not step in when questionable verbal comments were made. Instead, Amandine would suggest to students "yeah well, that's the job" and she now reflects negatively on her past actions:

> I really don't do that anymore, meaning, I really try to get the audio engineers feeling more self-confident, more clear about what they want, who they are, why they are doing this and making them decide where they're gonna put their boundaries. 'Cause I think it's my job to teach them that they can say no, that you know, this is not just the job.
>
> So those things will not happen in my classroom, bottom line, and if certain things happen, you know, we should be able to speak about it and everybody should be able to say how they feel about it. I don't wanna continue reproducing this lineage.
>
> (Amandine Pras interview, April-May, 2021)

Many colleagues that discussed EDI matters alluded to the greater influence of society on these challenges and that, not only was awareness the key, we as educators need to exemplify the change needed:

> These are three really complex things deeply imbedded into society.... We have a problem in the music industry, particularly the recording industry, we are not a viable option for young women who want to work here. There is absolutely no biological reason why women can't produce music.... Most if not all women in our lives have shown leadership and have shown direction and shown care and compassion. These are all things that you need to work in the recording industry, so the biological argument doesn't stand up for me. But I think that in the UK we are deeply wedded to this idea of genderisation of roles and those things happen before you are born. When I taught at a vocational college level, 16 to 18 year olds, half of my class were young women, so 15 women and 15 guys. And then I get to university and maybe five percent are women. So, something's happened in that point where women have made a decision, which is pretty complex, it's got various different emergent strands and elements that they're taking into consideration and they don't see that as a viable career? We need to do something about that.
>
> (Paul Thompson interview, April-May, 2021)

In my discussions with Jay Hodgson, he mentioned a theory where young men are being taught it's okay to fail and women are taught to be fearful of failure. Paul also discussed differences in societal upbringings for various genders:

> We know as men, we put our hands up for things that we weren't fully qualified to do, but we thought "I'll give it a go". But society discourages women from coming forward if they are not completely understanding of what they're talking about, which is why when women in those positions speak, men need to listen way more because they've gone through so much more to get to that point. Often, their point is much more considered and articulate than any guy at the same level, because they've had to get through so significantly more barriers. You did ask about race and gender, the things we can do, as white guys in the system, we can be part of the problem if we're not careful. Because we normalise particular language and particular examples of what is "musical" and who the practitioners are. When I wrote my book, I was careful not to roll out the same examples and so I included examples of practitioners and musicians that were women and people of colour.
>
> (Paul Thompson interview, April-May, 2021)

It becomes apparent that many of the non-white male educators I spoke with embody the action that is needed in the industry to normalise the change that is necessary. They are doing the work, making things happen and exemplifying that any person regardless of race, gender and social status can produce music or facilitate PMPP. It is important to be aware that they must be allowed to do so. The way through this problematic situation (in my opinion) is to promote awareness regarding equity, diversity and inclusiveness but manifest long term social and cultural change through the normalisation of our actions.

Higher education frameworks that support EDI

The ways that HE institutions facilitate music education need to be re-imagined so that diverse, inclusive and equitable environments are made available for all types of music and musicians, including popular music genres of all kinds. I understand there is only so much any one institution can do to accommodate the entire aspiring cohort, however, some collective change in this regard would (in my opinion) help to manifest the necessary cultural shift away from problematic EDI traditions of the past and any perceived hierarchy of music types. Within my institution change is taking place and much of what I present here has been taken from the rich discussions I have had with my colleagues.

Educational frameworks of the future should position students within various programs that are equally valued and appropriately funded, but foundational courses could include an integrated cohort (music theory, for example) that consists of students from all areas of the institution. This will help to develop interdisciplinary collaboration and an inclusive music educational culture within the institution, and minimise the genre-specific silos of the past. Programs should be flexible and allow students to choose and, if need be, adjust their pathway (major, specialisation) in accordance with their desires. Students should be able to select most of the music with which they work and some of the courses that they feel are best suited to their futures. Students should be empowered to unpack and develop their own sense of identity through their study and via mentorships as opposed to being moulded towards historical exemplars. Some recent scholarly work into this field suggested the key concepts of music education needs:

1. To raise consciousness of contemporary potential and challenges for musicians and HMEIs [higher music education institutions] in society, and how these interact with the complex responsibilities that HMEIs have, artistically, educationally, economically, socially and environmentally.
2. To encourage HMEIs to open their educational offer towards more diversity and to promote inclusiveness throughout their activities.
3. To embed entrepreneurial skills in the education of the artist to better prepare students for their future role as musician-entrepreneur.
4. To help music students and teachers internationalise their careers and activities…
5. To explore and discuss new Learning & Teaching models enabling HMEIs to educate creative and communicating musicians.
6. To encourage the use of digital technologies in music education.

7. To strengthen [the] student voice within the Association and within all AEC member institutions and to establish a European [and associated] network of HME students.
8. To increase the quality of early childhood music education and thus extend the audience of tomorrow.

(Association Europeénne des Conservatoires, 2021, p. 2)

I support these types of ideas and want to highlight a discussion I had with Alexis Kallio regarding music education and the necessary pathways forward for cultural change to occur. Alexis eloquently discussed the matter of equity in music education:

> I think that issues of equity are the biggest challenge, because how this is approached is multifaceted, context specific and complex. I think that popular music has often been framed as a democratic alternative to WAM that is 'automatically' more inclusive, accessible, participatory and so on. But there is a growing body of research (my own included) that has suggested this is not the case. I think, as popular music researchers and pedagogues, we need to move beyond straightforward questions of what and how to really consider why, and what for. These are big questions that pertain to who our students are, who we hope they become and what kind of society we want – both within and beyond music itself. I think that popular music, in particular, has potential to do great things in this space, but not if it assumes a critical pedagogical approach that reinstates the authority in the teacher, though perhaps not in content or skill, but in knowing where we all want to end up. This also demands intense theorisation within our field, which often is seen as divorced from practice or approached at a sort of surface level (as music scholars we are all playing 'catch up' with theory anyway, as we spent so long learning the craft of music-making, the craft of pedagogy, the craft of research etc) – so this, to me, makes sense as a starting point for asking big questions like those I posed above.
>
> (Alexis Kallio, questionnaire response, April-May, 2021)

FINAL WORD

As higher education facilitators, we must emphasise the potential that students see in their future. In my opinion, students must be allowed to lead the way, when they are ready, through their educational journey. If this is allowed to happen, and the educational experience epitomises much of the change that has been outlined as necessary throughout this book, then important societal cultural change is possible as a new breed of musicians filled with self-confidence, an appreciation for music of all types and innovative ideas enter the real world. These graduates would bring with them appropriate EDI practices as normalised agency as they move into the diverse fields that an education in music can support. Change may be slow to manifest at first, but as the years pass and graduates of this kind proliferate throughout the world, change would undoubtedly happen. Perhaps these music makers of the future will also develop new music industry business models that are far removed from the current inequitable system within which we operate. Imagine a music industry that considers all the ways in which music is consumed and appropriately distributes renumeration (wealth, notoriety and respect) to those who write, produce and perform music of all types. Imagine a world that values all music makers, from pop stars to street performers, as important citizens who play integral roles in society. Imagine the arts more generally, and arts education, being identified by the governments of the world as integral elements that nurture human existence and are supported as such. If this were to happen, and I do hope it is the case, all of us could then truly embrace and celebrate the joy of music making and what it means to us as human beings. Let us hope this is how it will be.

REFERENCES

Association Europeénne des Conservatoires (2021). *Strengthening Music in Society project outcomes.* Retrieved from https://aec-music.eu/publication/aec-strengthening-music-in-society-brochure/

Bridgestock, R. (2009). The graduate attributes we've overlooked: Enhancing graduate employability through career management skills. *Higher Education Research & Development, 28*(1), 31–44, doi: 10.1080/07294360802444347

Brooks, A., Pras, A., Elafros, A., & Lockett, M. (2021). Do we really want to keep the gate threshold that high? *Journal of the Audio Engineering Society, 69*(4), 238–260.

Eaton, K. (Producer). 2021. Hit the reset button: Rethinking how we teach music technology. Retrieved from http://www.instituteforcreativity.org/practice-magazine-article/hit-the-reset-button-rethinking-how-we-teach-music-technology

McIntyre, P., Fulton, J., Paton, E., Kerrigan, S., & Meany, M. (2018). *Educating for creativity within higher education: Integration of research into media practice.* London: Palgrave MacMillan.

Stewart, D. W. (2010). The purpose of university education. *The Psychologist-Manager Journal, 13*(4), 244–250. DOI: 10.1080/10887156.2010.522480

Appendix A

The doctoral case study

As a part of my doctoral research (Anthony, 2019), I conducted an ethnographic case study within an Australian University's Bachelor of Popular Music (BPM) program. The case study was centred around two third-year core courses, *Music Production 1* and *2*, that ran over two 12-week trimesters from February 2018 to October 2018.

THE BACHELOR OF POPULAR MUSIC PROGRAM (BPM)

The BPM was a stand-alone bachelor's degree in popular music. It was a non-traditional model for study in higher education in that it:

> seeks to emulate the learning practices of popular musicians in the community, unlearning the pedagogy of Conservatoire music-teaching and substituting a modified informal pedagogy in which the community of practice is at the centre of largely self-directed learning activity.
> (Carey et al., 2006, p. 4)

Students were selected for entry into the program through a rigorous interview and audition process (performance), and the cohort included students with strengths as songwriters, performers, musicians, programmers, engineers and producers. The program ran over a three-year model full-time, which included engagement in four core streams. These were (1) the history and analysis of popular music, (2) songwriting, (3) sound recording and music production and (4) major study – the composition and recording of original material. The BPM also integrated courses in basic music theory and ear training, the music industry, stagecraft, musicianship and a list of electives that included work placement opportunities that students could choose from to diversify their learning so it matched their intended industry integration.

Of particular relevance to the pedagogical approach of the BPM is that all of these courses were designed to support the peer learning process in a self-directed learning environment, and to develop diverse critical listening skills. In the program, a strong emphasis was placed on the composition, recording and performance of original material and students were encouraged to collaborate with other members of the cohort and musicians from outside the university to do this. Students were also given 24-hour access to a diverse range of instruments, recording studios, post-production rooms and a MIDI-lab.

Courses: Music Production 1 and 2

Music Production 1 and *2* were third-year courses in the music production stream of the BPM. Students who attend these courses had already received two years' education in sound recording, songwriting and the history of popular music. Students had also worked on their original material as part of their major study for two years and it is accurate to suggest that they brought some considerable prior learning in all of these areas. *Music Production 1* and *2* therefore brought together singer/songwriters, musicians and producer/engineers into the recording studio for a collaborative music-making experience where the ultimate goal was to engage in, and learn about, the agency of PMP. The courses were delivered via a lecture and tutorial framework wherein the content of the lecture informed the practice-based engagement in the tutorial. The tutorial work was mostly practice-based and equal emphasis was placed on the function of the song, musicality, the social environment and the use of technology during the recording process.

The various genres explored in the tutorial space were influenced by the musical preferences of the students. There was no limitation of musical genres placed on the students other than a stipulation that they must engage with music that utilised acoustic, electronic and online frameworks throughout the year. As such, the music content of the tutorials changes depending on the cohort of each iteration. In previous years, myriad genres were explored in the tutorial space, including Indie Rock, Christian Rock, Rap, Hip Hop, Metal, Pop, Folk, Country, Progressive Rock and EDM. This approach framed learning within the context of immersing students into specific music production cultures which all required diverse creative practice options. By doing so, these courses encouraged students to develop a greater understanding of PMP as a form of agency rather than learning *how to* produce a specific genre of music. Specific genre affordances were recognised and targeted by students, but this happened as a part of the music production cultures that the students explored. Students were also free to engage in any type of music making as a part of their autonomous work outside of class. This work was intended to focus student goals within their aspirations as musicians and the music was regularly critiqued in peer feedback sessions. Since 2015, a tracker assessment item has also been added to *Music Production 2* (see Anthony, 2020). This modern music-making experience facilitated further student engagement with electronic music and programming via online interaction as part of the pedagogical experience.

PEDAGOGICAL APPROACH

The lecture series of *Music Production 1* and *2* consisted of 24 face-to-face lectures that were delivered in a lecture room with PowerPoint presentations and included many videos and musical examples. The lecture content was designed to cover three main topics: (1) historical production approaches; (2) modern production approaches; and (3) overarching cultural, social, musical, cognitive and technological concepts that influence PMP as a form of agency. The lectures were delivered by the educator and often engaged student interactions via verbal questions, role playing, and peer-based discussions. There was a large amount of music played in these lectures; students learnt to use their critical listening skills weekly and then group discussions unpacked creative practice processes that influenced musical outcomes. The primary aim of the lectures was to give the students useful PMP knowledge that they could integrate into practice during the tutorial. The lecture content informed some of the practice in the tutorial and therefore was an important aspect, but it was not the focus of the research. The students' engagement within the tutorial was the priority.

The tutorials of *Music Production 1* and *2* allowed up to 15 students to enrol in each tutorial. These were delivered in a recording studio that consisted of a large control room and recording room. For the duration of *Music Production 1* and *2*, there was also a significant amount of work that occurred outside the tutorial experience in preparation for future classes. This work could be undertaken in the MIDI-lab, which was a room full of computers with various DAW software applications installed, or post-production rooms that were similar to small project studios. Also, students regularly took sessions home and worked on them in whatever home studio provisions they had. This approach was intended to engage students with the many locations that recording studios reside in the current professional industry: large recording environments, small studios, and home studios. It also engaged students with the varying technological affordances of those environments.

The tutorial activities were organised and framed for learning outcomes by the educator and a suitable array of production activities and creative practice experiences that students could engage with were planned. The actuality of the experience, however, was reactive to student engagement. This is where the more student-led processes came to fruition and dominated the learning process. The learning process was also outlined to students early on so they knew what they were expected to do as part of the tutorial experience. It was expressed to the students that, in the tutorial, profession-based music-making experiences were planned. The students then *do*, and then reflect on their practice and discuss their learning. Working with diverse forms of music enabled the tutorials to fuse concepts of PMP rather than commodities and allowed the students to engage with both hard and soft skills of PMP from within these tutorials.

The tutorial experience resembled a PMP learning culture where students were encouraged to "simultaneously learn producing and produce learning" (Bell, 2019, p. 175). In the tutorials, the specific activities utilised changed every week. Sometimes the students were playing music and capturing performances while some tutorials started with cognitive engagements that developed to complex feedback sessions regarding potential production approaches. Often, I led the tutorial with some very formalised instruction on how to do various creative practices. In later tutorials I intended for the students to completely drive the music-making process and the class often changed instinctively through many of these frameworks in the space of one tutorial.

In most tutorial situations artists, producers, musicians and the educator engaged in a PMP activity and then spent time reflecting on the day's work. This provided an opportunity for the educator and students to gather some useful information regarding engagement, what they found positive about the process, what areas they struggled with and how they reflected on their learning within these experiences.

THE FIVE PRIMARY PARTICIPANTS

Five primary participants took part in the semi-structured interviews and focus group discussions. They were selected from a list of volunteers from Tutorial 1 and it was intended that these participants represented the overarching student population on a smaller scale. Their individual details are provided below and, to protect the participants' identities, pseudonyms are used:

1. **Georgio** is a 27-year-old male who is a multi-instrumentalist (drums, bass, synthesisers, vocals) who also tours locally performing his own blend of electronic industrial music. He also regularly operates as a session musician for a variety of artists in performance and recording applications. Georgio is a producer, and in that role he engages with both electronic and acoustic-based production of his and other people's music.
2. **Elliot** is a 25-year-old male who is an international student and a multi-instrumentalist (keyboards, vocals) who performs his original material regularly. Elliot is a producer, and in that role he engages with both electronic and acoustic-based production of his and other people's music.
3. **Lennon** is a 20-year-old female singer-songwriter who is a multi-instrumentalist (piano, guitar). Lennon is a prolific songwriter who performs, writes and records her own music with other producers. She also engages in the production of songs she has co-written for other artists.
4. **Thom** is a 21-year-old male who is a multi-instrumentalist (drums, bass, electric guitar) who regularly operates as a session musician for a variety of artists in performance and recording applications. Thom is an aspiring sound engineer/mixer/producer, and in that role he engages with primarily acoustic-based production of his and other people's music.
5. **Hayley** is a 20-year-old female singer-songwriter (vocals). Hayley performs, writes and records her own music with other producers whilst being the lead singer in various bands. Haley is a prolific collaborative songwriter and also engages in the production of songs she has co-written for other artists.

The five primary participants represent a variety of popular musicians who engage with PMP and often do so within a HE context. They were all over the age of 20 years and have experience as popular musicians prior to university study.

GENERATION OF THEMES FROM DOCTORAL STUDY

The doctoral study collected much qualitative data through online surveys, semi-structured interviews, focus groups, a researcher's journal and video footage of the tutorial experience. There were many themes that were generated and triangulated and these have informed the questions that I posed to international students and educators as a part of the data collection for this book. The doctoral study themes were grouped together to generate some overarching domain clusters which can be seen in Table A.1.

Table A.1 Themes and generation of domain clusters

Themes	Domain Clusters
Intrinsic learning of music production hard skills from within diverse music production cultures	**1. The impact of diverse music production cultures.**
Understanding and learning the deeper and diverse connotations of communication, critical listening, engagement with emotional transference and psychological skills	
Developing soft skills	
Ensuring the correct balance of musicality and technological use to facilitate music production cultures that promote multi-skilled graduates	
The producer archetypes, the producer's vision, and diverse music production cultures	**2. Student accountability and engagement with the professional world.**
Engaging with profession-based activities and industry professionals in and out of the classroom	
Nurturing the vibe within a higher educational context	
Importance of production knowledge tradition to inform creativity	
Engaging with creativity and the subjectivity gap in many contexts – practical and cognitive.	
Dealing with conflict as part of creativity	
Accountability: individual, group, how you practise is how you play	
Formal instruction can be used to solidify student knowledge	**3. The educator's role**
The importance of class mode, feedback sessions, and monologues	
Turbocharged learning	
Student led frameworks and student ownership of the tutorial	
Developing relationships and breaking down the educator/student dynamic	
Collaboration increases learning productivity	**4. Learning occurs in many ways**
The students' social development as people, getting to know their strengths and weaknesses	
How students feel they learn and learning roadblocks	
Elements that promoted learning	
The importance of the lexicon used and the use of humour	
Making mistakes is ok	
Learning is implicit and tacit	
Learning and unlearning	
The shape of high-quality higher education music production cultures	
The development of life skills	

Following the completion of my doctoral research there have been many adaptations made to the pedagogical approach that I now promote for advanced PMP courses. These adaptations are represented throughout this book.

REFERENCES

Anthony, B. (2019). *Perspectives on learning popular music production in higher education from "both sides of the glass"* (Doctorate of Education Thesis, Griffith University).

Anthony, B. (2020). The tracker producer and top-liner collaborative music production project. In A. Bell (Ed.), *The music technology cookbook; ready-made recipes for the classroom* (pp. 275–282). New York: Oxford Scholarship.

Bell, A. P. (2019). Of trackers and top-liners: Learning producing and producing Learning. In Z. Moir, B. Powell, & G. D. Smith (Eds.), *The Bloomsbury handbook of popular music education: Perspectives and practices.* New York: Bloomsbury.

Carey, G., Draper, P., Lebler, D., & McWilliam, E. (2006). Learning and unlearning: New challenges for teaching in conservatoires. *Australian Journal of Music Education, 1.* Retrieved from http://hdl.handle.net/10072/13298

Appendix B

Useful lecture and tutorial content

Table A.2 presents some lecture and tutorial topics I have utilised in the past in my advanced PMP classes. The list is in no way representative of all the possibilities or in any particular order, but it exemplifies content that I have found useful to my students and their learning. The table also outlines the skills utilised by students in the related activities. PMP skill use: critical listening (CL), musicality MU), songwriting (SW), communication and psychological skills (CP), sound engineering (SE), engagement with emotional transference (EM) Programming (PG). Tutorial activity: cover (C) original (O)

Table A.2 Useful lecture and tutorial content

Lecture content	Tutorial activity	Tutorial learning outcomes and creative practice	PMP skill use	Pedagogical method
PMP skills, creative practice and cultures.	Live recording v overdubbed. (C)	How creative practice and technology affect the musical output. Four-hour session where students record the same song twice. One live and then one overdubbed using four overdubs if possible.	CL, MU, CP, SE	Formal, practice-based.
Record production of the 1920s, 30s & 40s.	Live recording practice. (C)	Capturing of a performance, using technology to do so. Understanding how creative practice influences the output. Musicians learning to balance in a room acoustically.	CL, MU, CP, SE	Practice-based, semi-student led.
Record production techniques of the 1950s-1970s.	Les Paul overdubbing project. (C)	Sound on sound overdubbing, historical recording, mixing in real time. Drums, bass, guitar, keys, lead and backing vocals. Analogue 2-track only.	CL, MU, CP, SE	Practice-based, semi-student led.

(*Continued*)

Table A.2 Useful lecture and tutorial content (Continued)

Lecture content	Tutorial activity	Tutorial learning outcomes and creative practice	PMP skill use	Pedagogical method
Record production techniques of the 1980s-1990.	Produce a track from start to finish (O)	Use a series of tutorials to produce a track of any genre to completion. Or to re-imagine a cover song in a contrasting style.	CL, MU, SW, CP, SE, EM, PG	Practice-based, semi-student led.
Multi-dimensional critical listening.	Live recording assessment workshop. (O)	Achieve a high-quality live recording (acoustic or electronic music) with video footage in four hours. Including professional session musicians	CL, MU, CP, SW, SE, EM	Practice-based, student-led.
Giving and receiving feedback (role plays).	Student song presentation. (O)	Engagement with the producer's vision and explore and expand students' aural libraries.	CL, MU, CP, SW, SE, EM	Formal, student-led.
Producer psychology: developing relationships.	Pre-production acoustic song. (O)	Students run an acoustic pre-production session. Engagement with the producer's vision. Producer archetypes. Aural library. Designing music production culture for recording.	CL, MU, CP, SW, EM	Semi-student-led.
PMP creativity, the aural library. Producer archetypes.	Programming beats and backing tracks (combining songwriting, pre-production and production). (O)	Found sounds, samples, beats, grooves, feels, synthesis and layering to promote top-line writing.	CL, MU, CP, SW, SE, EM, PG	Practice-based, student-led.
The producer's vision.	Synths overdubbing. (O)	Engage with the culture of synth overdubbing to develop parts.	CL, MU, CP, SW, SE, EM, PG	Practice-based, student-led.
Concepts of groove, feel and musicality.	Lead vocals. (O)	Engage with the culture of lead vocals and generate a performance from the singer. Vocal techniques focus (belt, head voice, etc.).	CL, MU, CP, SW, SE, EM, PG	Practice-based student-led.
R&B, Hip Hop, Rap production and cultures.	Mixing. (O)	Develop strategies to action mixing as a 'musical process' (not a technological one) and engage with the culture of mixing (small groups 2–3 students max).	CL, MU, CP, SW, SE, EM, PG	Formal, Practice-based, student-led.
PMP Maps (unpacking the intricacies of past and present hits).	The Tracker Process. (O)	Hold a creative meeting: top liner, tracker producer. Producer's vision. Aural library.	CL, MU, CP, SW, EM, PG	Student-led.

How society and culture influence PMP	The Producer's vision. (O)	Students play original demos and give feedback on potential music-making approaches. Producer's vision. Aural library.	CL, MU, CP, SW, EM	Formal, student-led.
Advanced electronic sonic processes. (layering, etc.).	Feedback on tracker songs. (O)	Students present their tracker production and provide feedback.	CL, MU, CP, SW, EM, PG	Student-led.
PMP technology and emotion.	Stacked backing vocal production.	Students engage in double tracking, harmonising, ghosting vocals and creating sonic effects with backing vocal techniques.	CL, MU, CP, SW, SE, EM	Practice-based, student-led.
3rd party mixing (advanced mixing). Mixing as a performance.	Third party mixing. Mixing as a performance. (O)	Students engage with mixing a track for a client. Client present (small groups).	CL, MU, CP, SW, SE, EM	Practice-based, student-led.
Creating the vibe: socio-cultural environments for PMP.	Vocal production, songwriting, tracker song. (O)	Students are required to listen to a semi-completed song, engage with the song on an emotional level and help students finalise vocal hooks.	CL, MU, CP, SW, EM, PG	Practice-based, student-led.
Vocal production.	Student completed song presentation (pre-mix). (O)	Students give feedback on their peer's songs (pre-mix). They highlight elements in the mix that could be accentuated, improved, or mix ideas that would enhance the production. This class can integrate an actioning stage, where new students act as third- party mixes to complete the work.	CL, MU, CP, SW, EM	Student-led.
Project management	Creative meetings with industry partners	Students undertake initial client liaison meetings with industry partners to develop a project that they will action in their autonomous work. Includes projects aims, scope, budgeting, deadline etc. (small groups).	CL, CP, SW, EM	Practice-based, student-led.
Commercial vs self-expressive works (release strategies).	Electronic music sonic effects. (O)	Students work on their electronic music projects applying side chain, autotune, warping, delay, distortion etc. To enhance sonic landscape of work.	CL, MU, CP, SW, SE, EM	Practice-based, student led.
Music producer entrepreneurial practices and industry theory.	Lead vocals 2.0 generating a performance on original song.	Students focus on social engagements (nurturing of a performer) that generate a performance from the singer.	CL, MU, CP, SW, SE, EM	Practice-based student-led.
Critical enquiry and PMP.	Collaborative top line writing.	Students work in small groups to write a top line with lyrics over prefab backing track.	CL, MU, CP, SW, SE, EM	Practice-based student-led.

Appendix C

International student survey questions

Sixty international student graduates that had studied popular music production as a part of their undergraduate degree programme provided responses to the following online survey.

1. What is your age?

   17–21
   22–25
   26–35
   36 and over

2. What is your gender?

   Man
   Woman
   Non-binary
   Prefer not to say

3. Where did you study popular music production at university?

   USA
   UK
   Europe
   Australia/New Zealand
   Asia
   South America
   Other

4. When did you study popular music production at university?

   before 2000
   2000–2010
   2011–2020
   2021

5. Please outline the degree program or course you studied Popular Music Production under. For example: Bachelor of Music (popular music major), Bachelor of Creative Industries etc.

6. Please select any roles you identified with prior to university study. Please select as many answers as suitable.

   Song Writer
   Sound engineer/mixer
   Performer/Musician
   Producer
   Music Programmer

7. Please select from the list below any descriptions that represent the rooms used for your popular music production classes. Check off as many options as you feel appropriate.

   Large scale recording studios with a control room and one or more recording rooms.
   Medium scale recording studios with a control room and recording room.
   Home/bedroom recording studio: one room with DAW and recording taking place in that room that resembled a home environment.
   Electronic music studio/Midi suite: a small room with a DAW, speakers and interface and equipped for electronic music production and perhaps vocal overdubbing.
   Online classrooms: all collaboration was facilitated online.

8. Rate the amount you engaged with 'doing' and 'learning' sound engineering and mixing during your classes that focused on music production.

   Never
   Some of the time
   50% of the time
   Most of the time
   All of the time

9. Rate the amount you engaged with 'doing' and 'learning' song writing during your classes that focused on music production.

   Never
   Some of the time
   50% of the time
   Most of the time
   All of the time

10. Rate the amount you engaged with ‘doing’ and ‘learning’ critical listening during your classes that focused on music production.

    Never
    Some of the time
    50% of the time
    Most of the time
    All of the time

11. Rate the amount you engaged with ‘doing’ and ‘learning’ music programming during your classes that focused on music production.

    Never
    Some of the time
    50% of the time
    Most of the time
    All of the time

12. Rate the amount you engaged with ‘doing’ and ‘learning’ communication and psychological skills during your classes that focused on music production.

    Never
    Some of the time
    50% of the time
    Most of the time
    All of the time

13. Rate the amount you engaged with ‘doing’ and ‘learning’ musicality and excellence in performance during your classes that focussed on music production.

    Never
    Some of the time
    50% of the time
    Most of the time
    All of the time

14. Rate the amount you engaged with ‘doing’ and ‘learning’ how to engage with emotional transference during your classes that focused on music production.

    Never
    Some of the time
    50% of the time
    Most of the time
    All of the time

15. How often did you work on acoustic recorded music in your popular music production course/classes/assessment

    Never
    Sometimes
    50% of the time
    Most of the time
    All of the time

16. How often did you work on electronic music in your popular music production course/classes/assessment

    Never
    Sometimes
    50% of the time
    Most of the time
    All of the time

17. In your popular music production course/classes/assessment, how often did you work on music that was generated through online collaboration only?

    Never
    Sometimes
    50% of the time
    Most of the time
    All of the time

18. Were your classes that specifically targeted popular music production more theoretical or practice-based? (Note: this question refers to advanced or final year classes, not classes on "how to do" sound engineering or use a DAW etc [skills-based classes]). Please target your answer toward any advanced classes you had that focussed on preparing you to work as music producers.

    All theoretical
    mostly theoretical
    50% theoretical and practice-based
    Mostly practice-based
    All practice-based

19. Were your classes that specifically targeted popular music production more teacher-led, student-led or a bit of both? (Note: this question refers to advanced or final year classes, not classes on "how to do" sound engineering or use a DAW etc [skills-based classes]). Student-led means that students ran the classes and took control of the activities

in the classes. The educator may have been present but students drove the class activities forward.

All teacher-led
Mostly teacher-led
50% teacher-led and student-led
Mostly student-led
All student-led

20. Whilst at University, did you work on the genres / types of music that you liked and listened to? Explain how the music you worked on was selected.

21. In your classes on music production, was music production creativity clearly explained to you? If so, please describe how ‘creativity’ was explained to you.

22. In your classes on music production, was your educator more of a teacher, facilitator/mentor, peer/music colleague or all of the above? Explain what you liked or did not like about your teacher-student experience.

23. In your classes on music production, did you engage with outside professional practitioners as a part of your studies? If yes please outline; if no, would you have liked that? In what form?

24. In your music production classes or within any assessment items, did it ever feel like you were doing something that was just like working in the industry? Please explain if either yes or no.

25. If you had any advice for the universities around the world regarding how they facilitate music production education what would it be? Please feel free to unpack any thoughts on classes, content, the educator etc. that are informed from your previous positive AND negative experiences.

Index

Note: **Bold** page numbers refer to tables; *italic* page numbers refer to figures and page numbers followed by "n" denote endnotes.

For Product Safety Concerns and Information please contact our EU representative GPSR@taylorandfrancis.com Taylor & Francis Verlag GmbH, Kaufingerstraße 24, 80331 München, Germany